NUTRITION AND THE WAR

NUTRITION
AND THE WAR

BY

GEOFFREY BOURNE, D.Sc.

Beit Memorial Research Fellow

CAMBRIDGE

AT THE UNIVERSITY PRESS

1940

CAMBRIDGE UNIVERSITY PRESS
Cambridge, New York, Melbourne, Madrid, Cape Town,
Singapore, São Paulo, Delhi, Tokyo, Mexico City

Cambridge University Press
The Edinburgh Building, Cambridge CB2 8RU, UK

Published in the United States of America by Cambridge University Press, New York

www.cambridge.org
Information on this title: www.cambridge.org/9780521237796

First published 1940
First paperback edition 2011

A catalogue record for this publication is available from the British Library

ISBN 978-0-521-23779-6 Paperback

To

MY LITTLE SON

"Learn first thoroughly the economy of the kitchen; the good and bad qualities of every common article of food, and the simplest and best modes of their preparation:"

RUSKIN

CONTENTS

PREFACE

This little book is based on a series of articles on nutrition which was published under a pseudonym in *Discovery* during 1938 and 1939. It is produced now with the object of helping the layman to understand the problems of nutrition at a time when such a knowledge will be very valuable to him, and to help him choose alternative foods when some items become too expensive to purchase or even impossible to obtain. When the invitation to write the book was first received I wondered what sort of practical information about food the average housewife wanted most. Some inquiries were made among housewives and others, and two things were uppermost in their minds. These were first, what minerals, vitamins, etc. their usual foods contain; and secondly, if they were unable to obtain a certain food, what they should eat in its stead in order not to miss anything vital from their diets. The food tables given in the last chapter of this book are designed to answer these queries.

The first chapters of the book are composed mainly of a general summary of our knowledge of nutrition together with a discussion of some possible nutritional repercussions of the war.

As far as possible only the elements of the subject have been given, and complicating factors such as the specific dynamic action of proteins have been deliberately excluded.

Naturally there is a great deal of information, both practical and theoretical, which cannot be included in a book of this size. It is kept small in order that it may be sold at a price which will enable it to reach the greatest number of people. If it plays even a small part in helping some of them to solve their nutritional problems during these very trying days, the book will have been worth while.

G. B.

SQUITCHEY LANE,
OXFORD
March 1940

ACKNOWLEDGEMENTS

I wish to express my indebtedness to a number of authors for the use I have made of their books (listed below). Direct quotations have been acknowledged, but if due recognition has in any case been inadvertently omitted I trust that the authors and their publishers will accept my apologies in advance. For the sake of space full details of the large number of scientific papers consulted is not given.

To Professor J. C. Drummond of the University of London, who, despite many urgent calls on his time, read critically the manuscript and gave me much helpful advice, I owe a considerable debt of gratitude. Professor Drummond, however, is in no way responsible for any statements or suggestions made in this book; these must be attributed solely to me.

Mr S. A. Barnett of the Dept. of Human Anatomy, Oxford, has given me a great deal of valuable assistance in "shaping" the manuscript, and I would like to thank him and my wife for carefully reading and checking the proofs. I am also indebted to Dr R. J. L. Allen of Cambridge for many helpful suggestions.

Finally I would like to thank Dr C. P. Snow of the University Press, Cambridge, for publishing in *Discovery* the articles on which this book is based, and the Cambridge University Press itself for the care taken in its preparation.

LIST OF BOOKS CONSULTED

Bacharach, A. L. *Science and Nutrition.* C. A. Watts and Co., 1938.

Clark, F. Le Gros and R. Titmuss. *Our Food Problem.* Penguin Special, 1939.

Drummond, J. C. and Anne Wilbraham. *The Englishman's Food.* Jonathan Cape, 1939.

Eddy, W. H. and G. Dalldorf. *The Avitaminoses.* Williams and Wilkins, 1937.

Harris, L. J. *Vitamins.* Cambridge University Press, 1937.

Harris, L. J. *Vitamins and Vitamin Deficiencies.* Vol. I. J. and A. Churchill, 1938.

Hutchison, R. and V. H. Mottram. *Food and the Principles of Dietetics.* Edward Arnold, 1933.

Mellanby, E. *Food in Health and Disease.* Oliver and Boyd, 1934.

Orr, J. B. *Food Health and Income.* Macmillan, 1937.

Plimmer, R. H. A. and V. Plimmer. *Food, Health, Vitamins.* Longmans, 1936.

Practitioners Handbook. "Diet in Health and Disease." Eyre and Spottiswoode, 1939.

Sherman, H. C. *The Chemistry of Food and Nutrition.* Macmillan, 1932.

Stern, F. *Applied Dietetics.* Williams and Wilkins, 1936.

CHAPTER I

Energy and Food

Do you think it is worth while bothering about your food in war time? Do you think diet is of any significance at all in your life? Perhaps you don't: but let me tell you of an experiment carried out in India by one of the pioneer workers in the modern science of nutrition— Sir Robert McCarrison.

When Sir Robert was working on nutritional problems in India he noticed that the Sikhs showed a remarkable degree of physical development. They were in perfect physical condition, with skins clear and shining; and according to Sir Robert "possessed powers of endurance which are probably unsurpassed by any other races of mankind". Upon investigation their diet was found to be composed mainly of coarsely ground whole wheat, milk, milk products, tubers, roots, green leafy vegetables, fruit, with only an occasional meal of meat.

Thinking of the white bread and margarine, jam and tea diet of so many of his fellow-countrymen at home, Sir Robert devised an experiment to compare this diet with the food eaten by the Sikhs. He took forty healthy young rats. Twenty he placed on the good diet, which was composed of milk, milk products, fruit and green leafy vegetables. The other twenty he placed on the bad diet, which was made up of white bread, a substitute for margarine, tinned meat, tinned jam, vegetables cooked with soda, and tea sweetened with sugar to which a little milk had been added. Apart from these differences in diet the two groups of rats were treated in exactly the same way: their cages were kept scrupulously clean and they were given plenty of direct sunlight. The two groups of rats were kept on these diets for six months. Since a rat lives for about two years this is equal to one-quarter of its lifetime and is equivalent to fifteen or eighteen years of a man's life.

At the end of the six months Sir Robert took stock of his rats. Of those on the good diet three were dead— one of pneumonia (due to the cage being inadvertently left in a draught during a cold spell), one due to an injury, and one for no obvious reason. Their organs were healthy in appearance and no signs of disease (apart from the lungs of the animal which died of pneumonia) could be seen in any of them. The surviving rats of this group were "well grown, sleek coated and active". The animals on the bad diet were "ill grown, poor coated, weakly and listless" and nine out of the original twenty were dead. Of these, three had grown so weak that they had been killed and eaten by the other rats. Three others had died from broncho-pneumonia, a disease which was extremely rare in the well-fed stock rat colony of 375 animals. All the rats which had died in this group showed signs of upset of intestinal function, and in a number of cases of definite intestinal disease. The digestive tract in all those examined was dilated and the walls were thin; there were signs of severe constipation, the intestine was collapsed and growths were present in the stomach. All the surviving rats in this group weighed considerably less than those on the good diet.

This experiment, illuminating as it is, does not tell the whole story; for rats can make their own vitamin C in their bodies, human beings cannot. So a human being on such a diet will suffer the many other troubles due to a chronic deficiency of vitamin C. We must remember also that the rats in this experiment lived in large airy cages and were given daily periods in the sunlight. Compare this mode of life with that of most human beings eating this poor diet. Many of them are occupants of slum areas, without regular exposure to sunlight. In addition, the rats on the poor diet had nothing to do. To earn these foods their human equivalents have to work for long hours at tiring jobs, and for half the year they may hardly see the sun. Add all these factors to Sir Robert McCarrison's results and you will probably find a clue to a great deal of the unnecessary amount of disease and suffering which afflict the mass of people of this country, and indeed of most countries.

Sir John Orr, in one of the most challenging books of recent years (*Food, Health and Income*, 1937), showed that probably half the people in this country cannot afford to buy a diet better than that given to Sir Robert McCarrison's second group of rats.

What of the result? When recruits were called for the war of 1914, an alarmingly high percentage was unfit for military service. We do not know the figures for this war, but the better nutrition of the last twenty years will no doubt result in some improvement. The problem of satisfactory nutrition is mainly economic and can be met only in part by the education of the people to buy the right things.

Now if you feel that a knowledge of diet is of importance to you and your country, let us pass to an examination of our foods and what they contain.

We obtain our food from two sources. First, from the animal kingdom, and secondly, from the plant kingdom. There is considerable disagreement between countries and races, and even between individuals, as to what constitutes food and what does not. Most of us in this country, for example, do not fancy the snails which are dear to the French palate. We look with dismay at the cooked octopus served by some Soho restaurants, and we abhor the thought of consuming the grubs and moths so sought after by the Australian aboriginal. These variations are, however, nothing more than variations in taste. All these things are food and some of them probably very good foods. Then many of us regard tea, coffee and thin chicken soup as foods, whereas they are not foods at all. The first two are stimulants and the last merely flavoured salty water.

Few of us pause to consider the quantity of food which our digestive tracts handle in the course of a year. It has been estimated that a person on an adequate income consumes something in the neighbourhood of three-quarters of a ton of food in this time. A family of four will, then, dispose of two and a half to three tons of food a year. If everyone in the country were able to buy enough food the actual amount consumed would be something like thirty-eight million tons. This figure does not of course include those parts of the food which are not edible, nor those edible parts which are wasted in the preparation of the food for the table, nor food which is wasted in distribution.

Some of us for religious or other reasons regard a vegetarian diet as perfect, and it is frequently assumed that this is quite a modern food fad. Actually it is not, for the famous Greek mathematician Pythagoras is believed to have founded vegetarianism in the sixth century before Christ. There would be no good purpose served in this short book in joining the controversy of vegetarianism versus mixed diet. Both sides have good arguments. While most nutritionists of the present day favour the inclusion of animal products in a diet, there is the example of Paavo Nurmi, one of the most brilliant long distance runners the world has ever seen, who was a vegetarian. One point worth mentioning, however, is that most vegetarians permit themselves to eat eggs, butter and cheese and to drink milk, and are, therefore, not true vegetarians at all.

This brings us to the discussion of what function or functions our food fulfils in our bodies. It is required to do two main things. First, it must supply us with energy; secondly, it must build up the body in the growing stage and replace worn-out tissues when the actual growth period is finished. Practically everything you do requires energy for its performance; even mere existence necessitates the expenditure of some energy to keep the body "ticking over". I say *practically* everything you do, because I refer to things which require physical exertion. Mental work alone has not been found to cause any demonstrable expenditure of energy. Most forms of mental work, however, entail some form of physical movement, however small, and thus there is an apparent need for a little more energy than in absolute rest. A man lying motionless in bed, however, and doing difficult calculations in his head, cannot be shown to use any more energy than if he were lying there with his mind more or less blank. Existence without any form of physical exertion requires, as I mentioned before, the expenditure of energy, and the amount of energy required for this purpose is roughly the same for persons of the same weight, age, sex and state of health. For every form of physical activity the body expends energy over and above the basal amount just mentioned.

The energy required for living and working may be obtained from two sources. First, from our food, and secondly, by burning up our own tissues if food is withheld. If we are engaged in work which requires for its performance more energy than is contained in our food, then the balance of energy is obtained by burning up the body tissues, a state of affairs which cannot of course be permitted to go on for long without very serious results.

The human body in its relationship to energy has often been likened to a steam engine or a petrol engine. It differs from these in two fundamental ways: in being able to burn up its own substance in order to keep going in the event of a shortage of fuel, and in being able, if there is enough fuel, to replace its parts when they become worn. Otherwise the body might well be compared to a petrol engine. In both a fuel is required, in both the burning of that fuel results in the production of energy and heat. In our bodies we have elaborate mechanisms for regulating that body heat and so to make us independent of the surrounding temperature. Man belongs to a group of animals known as the mammals, one of the features of which is this ability to regulate the body temperature. Mammals include such beasts as monkeys, apes, lions, tigers, dogs, cats, horses, elephants, squirrels, kangaroos, and many others. Birds are the only other group of animals able to regulate the body temperature in this way. All other animal groups, such as those which include the snakes, lizards, crocodiles, turtles, frogs, fish, and so on, do not have this ability. In the case of fish, distinction must be drawn between the true fish, "Pisces", and the fish-like creatures, whale and porpoise, which are really mammals and which are capable of regulating their temperature independently of their environment, in common with other mammals.

To return to the subject of food. All engines which burn fuel to supply energy burn it at a high temperature. Substances used for human food could be burnt under a steam engine and would supply energy to drive that engine, but they would be burnt at a temperature which our own bodies could never withstand. In our bodies they are burnt at body temperature; that is, at about 37 degrees centigrade. This is accomplished by the aid of three principal types of catalysts—substances which facilitate a chemical reaction, but which can be obtained unchanged at the conclusion of the reaction. This definition cannot be applied completely to the catalysts of the body, because they are used up to varying degrees in promoting the chemical reactions which take place there. The three principal types of body catalysts are enzymes, vitamins and hormones. A distinction between these three substances cannot always be drawn. Enzymes are chemical substances formed in the various cells of the body. Vitamins are taken into our bodies with our food and are afterwards taken up by the cells. The term hormone is usually applied to chemical substances which are formed in the body by special organs known as ductless glands. As a result of these various catalysts the food is burnt at a low temperature and provides us with body heat and energy. The constituents of our diet which directly produce the heat and energy are the proteins, the fats and the carbohydrates. In addition to these substances the food must contain

vitamins, mineral substances, water and roughage.
Vitamins and minerals will be dealt with more fully in
subsequent chapters, but it may be as well to say a few
words here about the water and roughage in food.

Our bodies contain a large amount (about 70 per cent)
of water. We daily excrete large amounts of water, as
sweat (sometimes not recognised as such and spoken of
as insensible sweat), in our breath, in our urine and in
our faeces. Therefore we must consume water to make
up for this loss, and we obtain our water as such, as a
flavoured drink, or as a part of our food. Most fresh
food contains rather surprising amounts of water
(90 per cent in the turnip, 60 to 70 per cent in fresh
meat, 30 to 40 per cent in fresh bread).

The body also requires a certain amount of material
which is not digested and absorbed, in order to supply
bulk to the food. This material is known as roughage.
It is usually some form of cellulose, and is obtained from
vegetables and fruits and wholemeal bread. It is
believed to aid in maintaining the muscular "tone" of
the intestine and in that way to help in moving the food
along. Therefore a certain amount of roughage in the
food is essential for the proper working of the stomach
and intestines. This necessity for bulk in food is one
reason why we are not likely to have all our food
requirements reduced to a small pill, to be taken with
a glass of water instead of having a good square meal.

To return to the subject of energy. When coal or
petrol is burnt in an engine, oxygen is consumed, carbon
dioxide is given off and various unburnt or partially
burnt residues are left. In the same way the human
body requires oxygen to burn up its food, and one of
the products of this combustion is carbon dioxide. By

a measurement of the oxygen consumption and the carbon dioxide given off, a direct estimate of the energy requirements for various tasks can be obtained. For this purpose large chambers called respiration calorimeters have been built. In them men or women can live and perform various tasks. The chambers are built large enough to take a bed, table and chair, and a fixed bicycle or other device for expending energy. All oxygen passing into these chambers and all carbon dioxide coming out is measured, and, as the person inside sleeps, eats, writes or "rides" his fixed bicycle, so his energy requirement for this type of activity can be calculated. The amount of energy which is given off as heat is also measured. Drs Atwater and Benedict in America, some years ago, built a famous calorimeter of this type, with which they made a large number of measurements of the energy expended by men, women and children engaged in various tasks.

The energy value of a food is measured in calories. These nutritional calories are one thousand times greater than the calories familiar to students of elementary physics. This latter unit is the amount of heat required to raise 1 gram of water 1 degree centigrade. The term calorie as used in the study of nutrition means the amount of heat required to raise 1 kilogram of water 1 degree centigrade.

The energy value, that is, the calorific value of a food, is derived from its proteins, fats and carbohydrates. These substances will be described more fully in the next chapter.

The energy required to be supplied by the food depends primarily upon the type of work a person is called upon to do. It is also modified by other factors.

Generally speaking a woman is regarded as requiring only four-fifths of the calories needed by a man. This figure cannot be rigidly adhered to, because there are obviously many women whose calorie requirements are equal to or greater than those of many men. The energy requirement also varies considerably with age, a new-born baby requiring less total energy than an adult man or woman. Individuals themselves, even in health, vary to a certain extent in their calorie requirement, and in certain diseases the requirement will vary even more.

In a cold climate or season there will be a greater loss of heat to the surrounding air; consequently more food will be required to make good this heat loss. In a hot climate or season the air temperature may be hotter than that of the body, and there will be less food required because there will be no heat loss for which to compensate. The actual body temperature in a hot climate or temperature cannot, however, be appreciably reduced by reducing the amount of food eaten; but it can be reduced to a certain extent by eating foods containing more water in order to cause greater sweating, and by reducing the amount of protein eaten.

The energy requirement of a man lying in bed and kept at an equable temperature has been calculated. If he does nothing but this for 24 hours he will require 1700 to 2000 calories. Every movement he makes will require more energy for its performance. If he sits for 6 hours he will require about an additional 170 calories. If he does 6 hours' slow walking he will require a further 400 calories, and if he does 6 hours' moderate exercise he needs another 600 calories.

A number of estimates have been made of the calorie requirements for different occupations and the following list has been compiled from various authorities. For most of them I am indebted to the list published by Hutchison and Mottram in *Food and the Principles of Dietetics*:

Monk in cloister	2304 calories
Teacher or office clerk	2600 ,,
Tailor	2750 ,,
Weaver	2700 ,,
Physician	2762 ,,
Housewife, hand seamstress, or typist	2800 ,,
Soldier (peace, light work)	3029 ,,
Bookbinder	3100 ,,
Soldier (war, moderate work) ...	3146 ,,
Shoemaker	3150 ,,
Carpenter	3194 ,,
Metalworker	3500 ,,
Painter	3600 ,,
Royal Engineers (moderate work) ...	3818 ,,
Farm labourers	4100 ,,
University boat crew (each requires)	4085 ,,
Blacksmith	4117 ,,
Brickmaker (estimate for Italians) ...	4641 ,,
Stonemason	4850 ,,
Excavator	5000 ,,
Woodcutter	5500 ,,
College football team (U.S.A.) ...	5742 ,,
Brickmaker (American estimate) ...	8848 ,,

Dr Hutchison and Prof. Mottram also quote the case of a competitor in a 24 hours' cycle race in the U.S.A. who consumed food equivalent to 9000 calories during the race and also lost body substance equal to 1000 calories, so that his calorie consumption during that period was about 10,000 calories.

Prof. Mary Schwartz Rose of the U.S.A. has published a very useful table of the calorie requirements of various activities:

Form of activity	Calories used per hour
Sleeping	65
Awake lying still	77
Sitting at rest	100
Reading aloud	105
Standing relaxed	105
Hand sewing	111
Standing at attention	115
Knitting	116
Dressing and undressing	118
Singing	122
Tailoring	135
Typewriting rapidly	140
Ironing (5 lb. iron)	144
Dish washing (crockery)	144
Sweeping bare floor	169
Bookbinding	170
Light exercise	170
Shoemaking	180
Walking slowly	200
Carpentry, metalworking, industrial painting	240
Active exercise	290
Walking moderately fast	300
Stoneworking	400
Severe exercise	450
Sawing wood	480
Swimming	500
Running slowly	570
Very severe exercise	600
Very fast walking	650

From this list it is possible to work out for yourself just how many calories you need for a day. Multiply the number of hours for which you are engaged in various forms of activity by the number of calories required per hour for the performance of that activity and add the results. The figure you obtain will be your daily calorie requirement. For example, if you are a housewife you will probably need the following calories:

Sleeping, 8 hours	...	...	...	520 calories
Awake lying still, 1 hour	...	...	77	,,
Sitting at rest, 3 hours	...	...	300*	,,
Hand sewing, 1 hour...	...	...	111	,,
Dressing and undressing, ½ hour	...	59	,,
Ironing, ½ hour	...	...	...	72	,,
Dish washing, 1½ hours	...	...	216	,,
Sweeping, ½ hour	...	...	...	85	,,
Washing clothes, 1 hour	...	...	144†	,,
Walking slowly, 2 hours	...	...	400‡	,,
Standing around cooking, preparing food, etc., 3 hours ...	...	...	345§	,,
Knitting and darning, 2 hours	...	232	,,

Total, 24 hours, 2561 calories.

* This figure includes sitting at meals, etc.
† Probably the same as ironing.
‡ Including dusting, laying table, tidying, etc.
§ Taken as equivalent to standing at attention.

This result is, for a number of reasons, very approximate, but it gives some idea of a housewife's calorie requirement.

Children need fewer calories than adults. The number depends upon the weight and, therefore, roughly upon the age of the child. But as children expend energy at a greater rate than adults they need more food (i.e. more calories) in proportion to their weight.

A new-born baby weighing about 7 lb. and sleeping quietly requires about 6⅘ calories per hour; this would be equivalent to 153 calories in 24 hours, if it slept the whole time. Babies of course do not sleep through the 24 hours; they frequently spend what seem to be excessively lengthy periods in crying and during these times their energy output is increased by 100 per cent.

Cathcart and Murray, in a Medical Research Council Report published in 1931, give a list of the calorie requirements for children in Great Britain:

Years	Calories per 24 hours	Years	Calories per 24 hours
0–1	600	6–8	1800
1–2	900	8–10	2100
2–3	1200	10–12	2400
3–6	1500	12–14	2700

Up to the age of 14, boys and girls are assumed to require approximately the same number of calories, but Holt and Falls, in an investigation of children in New York, found that boys take in more calories than girls at a much earlier age. The list shown below is from the paper published by them in the *American Journal of Diseases of Children* for 1921:

Age (years)	Calorie intake (boys)	Calorie intake (girls)
1	950	940
2	1135	1110
3	1275	1230
4	1380	1300
5	1490	1410
6	1600	1520
7	1745	1660
8	1920	1815
9	2110	1990
10	2330	2195
11	2510	2520
12	2735	2860
13	3040	3210
14	3400	3330
15	3855	3235
16	4090	3160
17	3945	3060
18	3730	2950
Adult	3265	2640

The significant fact about these figures, and one which should be borne in mind by those who are concerned in fixing allowances for children, is that the calorie requirements of boys from 14 to 18 years of age and of girls from 12 to 18 are greater than those of adult men and women.

Old persons, in view of their lessened activity, need fewer calories than younger adults. One is faced, however, with the problem of "When is a man old?" Some persons, though old in years, are often found to be more active than other much younger men.

Without an adequate supply of calories the body is forced to use up its own substance to supply energy and one of the difficulties in war time is ensuring that all persons have the means to obtain sufficient food to supply, at the very least, enough calories.

Unfortunately many persons in this country are in receipt of incomes that enable them to supply only just enough calories for their families, but this is being made more difficult by the rising war-time prices. In a few cases wages have been raised in an attempt to meet this increase, but we have recently been warned that it will not be possible for them to continue to rise. Where a man is in receipt of an income adequate to supply ample calories for himself and his family, despite even a considerable rise in prices, this reduced buying power will mean simply a reduction in some of the comforts of life. But as mentioned above, many people in this country were not in this position even in peace time.* If prices so far outstrip wages that all these families are forced to reduce their food intakes, the effectiveness of the war effort of the nation is going to be radically reduced. Many men are being faced not only with the limitation of the buying power of their wages, but also

* According to Sir John Orr (*Food, Health and Income*, 1937) four and a half million people in this country were, before the war, in receipt of such tiny incomes that their diets must have been deficient in every known dietary essential.

with the demand for much longer hours of work in the interests of the country. If they are engaged in metal or carpentering work, as many of those associated with armaments production probably are, each one will require 140 more calories for each extra hour worked than if he were resting at home. Four hours' overtime will call for 540 more calories. If such a man is already living at the limit of his income, where are the extra calories to come from? If he cannot buy the necessary food he can only obtain them by burning up his own tissues, and this means, in effect, that he must inevitably break down. Such a state of affairs in a large number of men engaged in industry would rapidly jeopardise our war effort. Even if the undernourishment is not excessive, the resulting inefficiency of the work is, through no fault of the worker, going to be very costly.

This condition could be prevented if every man engaged in industry were supplied, without cost— by the Government if the firm is not prepared to do so—with a glass of milk and two or three slices of wholemeal bread and butter and cheese, together with a piece of fruit if possible, every day.* This would benefit not only the men, but also their families, since there would be more money available for their food if father had part of his calorie requirements provided at work. If this scheme were adopted the speed of production and the general industrial efficiency would be considerably increased, to say nothing of the improvement in the worker himself.

* I understand that some firms run canteens where cheap and nourishing meals may be obtained, and where milk may be bought cheaply. This is better than nothing, but it would be much better if the food were free.

CHAPTER II

Proteins, Fats and Carbohydrates

In the previous chapter it was shown that foods supply
energy to the body and that the amount of energy
required depends upon the work performed. In this
chapter it is proposed to discuss the substances in the
food that produce the energy—the proteins, fats and
carbohydrates.

PROTEINS

Proteins form the greatest bulk of animal tissues. The
characteristic of all proteins is that they contain nitrogen.
Carbon, hydrogen and oxygen are common to all of
them as well, and they may contain sulphur and
phosphorus. These chemical elements are combined to
form amino acids, and a protein is made up of groups
of amino acids. Typical of these acids are glycine,
alanine and tyrosine. There are many amino acids,
and practically any combination of any number of
them is possible; therefore it is not surprising that
there is a great variety of proteins. So great is this
variety that not only are the proteins of different tissues
different from each other, but, in addition, the proteins
of the same tissue in different species of animals are
different. For example, liver protein of man is different
from liver protein of sheep or rabbit.

When proteins are eaten or ingested they are broken
down principally by enzymes known as "proteolytic"

enzymes. They are first broken into large groups of amino acids, then into smaller groups, and then into very small groups or into individual amino acids. In this condition they are absorbed; they are carried away in the blood stream to the tissues and there they are reconstituted into proteins. The relationship of proteins to amino acids can be explained by comparing them to a house and its component bricks. Let us assume that we have a row of houses on one side of the road. The road is equivalent to the barrier between the protein in the food as it lies in the intestine and the tissues of the body. Each house represents one protein. Each of its constituent bricks represents one amino acid. We wish to move the houses from one side of the road to the other. First of all we take down the walls. This is equivalent to breaking the protein up into large "chunks" of amino acids. Then the walls are broken up into smaller groups of bricks and finally into the constituent bricks, each equivalent to a single amino acid. The bricks are carried by hand either singly or in small groups across the road, and there they may be rebuilt into houses. They may be used to build a house quite different from the original house. There are, however, certain bricks essential for the foundation, and there are facings which are necessary for every house. In the same way there are essential amino acids which must be included in the diet if satisfactory growth and replacement of tissues is to take place.

The amino acids are absorbed from the intestine into a large vein known as the portal vein, which carries them to the liver; from the liver they reach the body circulation and the various body tissues. From the variety of amino acids circulating in the blood each tissue selects those which it requires to replace or add to its own protein material.

Proteins are known as complete or incomplete. They are complete if they contain all the amino acids essential for maintaining life and permitting normal growth. They are incomplete if they lack any of these essential amino acids. Examples of complete proteins are lactalbumen found in milk, ovalbumen and ovovitellin of the egg, glycinin of the soya bean and excelsin of the Brazil nut. Two typical incomplete proteins are zein, the chief protein of maize, and gelatin. There are some points in connection with the latter that are worth mentioning. Gelatin is the constituent of certain soups that causes them to set as a jelly when cold. This is frequently taken to indicate that the particular soup is highly nutritious, quite incorrectly, since in the first place relatively little gelatin is needed in a soup to make it set as a soft jelly, and in the second gelatin is probably the poorest of all proteins. A much more nutritious type of soup would be one made with milk and peas, beans or lentils.

Nutritionists have generally come to the conclusion that those proteins which have come from animals are more economically used in building up the body tissues than those of plants, and they recommend a minimum daily intake of proteins derived from animals. The British Medical Association committee on nutrition recommends 50 grams of animal protein per day. Others have stated that a minimum of 50 grams of mixed animal and plant protein is essential, and the amount should preferably be more than this for the proper maintenance of health.

The daily protein requirements of children from 1 to

2 years are 35 to 50 grams. This amount increases steadily with age until children of 16 to 17 years of age require 150 to 160 grams of protein per day. The League of Nations *Report on the physiological basis of nutrition,* 1936, recommends that every person should consume 1 gram (one-thirtieth of an ounce) of protein per day for every 2 lb. (approximately) of body weight, so that a 10 stone man would require 70 grams or $2\frac{1}{3}$ oz.* of protein every day. The report considers that some of this protein should come from animal sources, or, as the writers of the report put it, that it should be "first-class" protein, as distinct from that of plants, which is described as "second-class" protein. The report urges that, during the growing period, animal or first-class protein should form a large proportion of the total protein taken. The following are the recommendations regarding protein intake which are published by the Nutrition Committee of the League of Nations:

Age (years)	Grams required per 2 lb. (approx.) of body weight
1–3	3·5
3–5	3·0
5–12	2·5
12–15	2·5
15–17	2·0
17–21	1·5
21 and upwards	1·0
Women:	
Pregnant. 0–3 months	1·0
4–9 months	1·5
Nursing	2·0

Every gram of protein burned in the body gives on the average 4·1 calories.

* Equivalent to about half a pound of lean steak.

FATS

Most of us are familiar with the substances which are included in the term fat. Liquid fats are known as oils, although not all oils are necessarily fats. True fats are substances which are composed of a fatty acid united chemically to glycerol. There are of course vegetable and animal fats, just as there are vegetable and animal proteins. Vegetables and plants vary a great deal in the amount of fat they contain. Nuts and soya beans contain relatively large amounts of fat. So does cocoa, which is made from the cocoa bean. Most animal foods contain fat, some of them very large amounts. Two typical examples are butter and cheese. The fat content of meat varies considerably with the cut.

When fats are eaten they are broken up into the fatty acids and glycerol of which they are composed. They pass through the intestinal wall in this condition and combine to form fat again. The fat is then taken up by the lymph vessels, and finally discharged with the lymph into the blood. The little particles of fat in the blood are known as "chylomicrons". They appear in large numbers in the blood after a meal during which large quantities of fat have been eaten. The fat is absorbed very rapidly by the tissues and a few hours after the meal most of it has disappeared from the blood.

The fatty acids which compose the fats are of two types; one sort forms liquid fats and the other solid. Oleic acid is typical of the first type of fatty acid and palmitic and stearic acids of the second. The last two are solid at ordinary temperatures, the first is liquid. In view of the fact that fats can be formed in the body from carbohydrates it has been thought that it is not essential for fats to be included in the diet, apart from the fact that they have a very high calorific value. It has been shown, however, that there are certain fatty acids which are (at least for rats) essential for normal health. Rats reared on a diet devoid of fat but containing all the other elements of a perfect diet do not thrive but develop a characteristic deficiency disease. The administration of linoleic or linolenic acid results in a cure. Other unsaturated or saturated fatty acids are ineffective. One does not know just how far these results are applicable to human beings, but the fact that this condition is possible in rats should make us take care to see that our diet contains reasonable quantities of fat. It is in any case advisable that a certain amount of the fat contained in our food should be of animal origin, because most types of animal fat contain some vitamins A and D, whereas vegetable oils and fats generally contain very little or none of either of these vitamins.

In the body, fat is an economical form of stored energy and also serves as a packing for various organs, such as the kidney. Animals which live on plants usually contain harder fat than those animals which live on meat. Warm-blooded animals also have much harder fat than cold-blooded animals. In ourselves and other warm-blooded animals the fats with the lowest melting point (the softer fats) are found nearest the skin. Fats with higher melting points (the harder fats) are found in the abdominal cavity. Fats form portions of substances, known as phospholipoids, which are important constituents of the membranes of living cells and form a large proportion of brain tissue.

When 1 gram of fat is burnt in the body it gives on the average 9·3 calories.

CARBOHYDRATES

Carbohydrates may be loosely divided into two forms—sugars and starches. They are more accurately divided into

1. Monosaccharides.⎫
2. Disaccharides. ⎬ Sugars.
3. Polysaccharides.   Starch and glycogen.

Carbohydrates contain only carbon, hydrogen and oxygen; they are the cheapest forms of energy available to man and they form the greatest part of the calorie section of his food. Carbohydrates are synthesised in the green parts of plants in the presence of sunlight. It is believed that in their formation water and carbon dioxide first combine to form a substance known as formaldehyde; this is poisonous to the plant and is rapidly changed into sugars. Sugars are stored up in the daytime and at night the plant sets about changing them into starch.

Monosaccharides. Typical of this group of carbohydrates is glucose, also known as dextrose or grape sugar. Glucose is present in the blood of all animals, in ripe fruits and in most plant juices. It is an important constituent of grapes; large amounts (relatively speaking) are also found in sweet corn. All other carbohydrates are reduced to monosaccharides during digestion, and are then combined with phosphorus and absorbed into the blood stream. The glucose of the blood is all the time being rapidly oxidised, that is to say, it is being burnt; and, just as rapidly, it is being replaced. It is estimated, according to the American nutritionist Sherman, that over half the energy manifested by the human body comes from the oxidation of glucose. Excess glucose is stored in the liver after being converted into glycogen. Another monosaccharide is fructose, which is also spoken of as laevulose or fruit sugar. In nature it usually occurs associated with glucose. It forms a considerable proportion of honey (50 per cent of the solid matter). Galactose, a third monosaccharide, is formed in the intestine from the digestion of milk sugar (lactose).

Disaccharides. The best-known disaccharides are cane sugar and beet sugar. Although they have different origins they have an identical chemical composition, and are both known scientifically as sucrose. Sucrose mixed with glucose is found in beets, sugar cane, sugar palm and sugar maple. Fifty per cent of the solid matter of pineapples is said to be sucrose. Upon digestion sucrose is broken up into the monosaccharides glucose and fructose.

Another disaccharide is lactose or milk sugar. Lactose forms 4 or 5 per cent of cow's milk and 6 to 7 per cent of human milk. When lactose is digested it forms glucose and galactose. A third disaccharide is maltose, which is formed from starch by the action of saliva or by the action of an enzyme derived from the pancreas. Later on in the process of digestion the maltose is broken up into glucose.

Polysaccharides. Starch is the principal polysaccharide. It is a food storage product of plants and is found particularly in seeds, roots and tubers, cereal grains and leaves. 70 to 80 per cent of the solid matter of potatoes is starch. It is also present in unripe fruit. Bananas contain a large quantity of it. When fruit containing starch ripens, the starch changes into sugar. The action of digestion on starch is to change it into sugars. If starch is heated another polysaccharide, dextrin, is formed, and this forms the sugar maltose on digestion. A third type of polysaccharide is glycogen, which is formed in the body from sugar. It is stored chiefly in the liver and is spoken of as animal starch.

All carbohydrates when taken into the body as food are converted by digestion into monosaccharides and absorbed into the blood stream as such. Upon absorption they are conveyed via the portal vein to the liver, where most of them are stored; from the liver they are released into the general circulation in small amounts as needed, and are then burnt to supply heat and energy. If unusually large amounts of carbohydrate are eaten the excess is changed into fat and stored in the tissues.

One gram of carbohydrate (sugar or starch) when burnt in the body gives on the average 4·1 calories.

SUMMARY OF ENERGY ELEMENTS IN THE FOOD

The energy elements in the food are the proteins, fats and carbohydrates:

Proteins are used to replace lost tissues and to provide new tissue and they may be burnt as fuel. One gram of protein yields 4·1 calories.

Fats may be oxidised directly to give energy or may be stored as a reserve supply of energy, or may be used for "cushioning" organs and tissues. One gram of fat yields 9·3 calories.

Carbohydrates (sugars and starches) may be oxidised directly to give energy or may be stored as glycogen. Excess may be changed into fat and stored in the tissues. One gram of carbohydrate yields 4·1 calories.

CHAPTER III

Vitamins

In the year 1888, Dr Lunin, who was carrying out research work on diet at Basle University in Switzerland, fed a series of mice on an artificial diet composed of pure fat, highly refined protein, carbohydrates, mineral salts and water. To his surprise he found that his animals rapidly sickened and died. Not long after this a Dutch physiologist, Dr Pekelharing, found that if he added a small quantity of milk to this diet, an amount of milk quite insufficient to cause any appreciable increase in the protein, fat, carbohydrate or minerals, the animals did not die. It was, however, left to Sir F. Gowland Hopkins of Cambridge to make a thorough investigation of the problem, and finally to announce in 1912 that a number of foodstuffs contained substances which he described as "accessory food factors". In 1913 Dr Funk, a Polish biochemist, who was working in England, described these substances as "vitamines". With these investigations there began a new dietary epoch which is not yet concluded. The "e" is now left out of the word vitamines, but it is still pronounced with the first "i" as in vital.

At first it was thought that there was only one vitamin, but it was quickly realised that there were several. After passing through a period of considerable confusion biochemists finally decided that there are five—vitamins A, B, C, D and E. Very soon, however, it was realised that vitamin B was not as simple as it had appeared at first sight: various workers began to separate out vitamin B_1, vitamin B_2, etc., until at the present day there are, according to the American nutritionist Prof. Elvehjem, something like ten vitamins in the B group. There have been a number of other vitamins added to the main list within recent years, and the "accessory food factor" of thirty years ago has multiplied rather alarmingly.

There are three classical diseases associated with vitamin deficiency. These are beri-beri, scurvy and rickets. These diseases are now known to be due respectively to lack of vitamins B_1, C and D.

Beri-beri is a disease of the nervous system in which various forms of paralysis accompanied by wasting occur. According to Dr L. J. Harris, in his excellent little book *Vitamins*, it was known in China as long ago as 2600 B.C. Beri-beri is a disease found particularly in those countries where the main diet is polished rice.

Scurvy was well known in England in earlier times, and at one period it was so common in London as to be called the "London disease". It is a disease characterised by bleeding gums and leakages of blood from blood-vessels into the surrounding tissues; the blood forms bruise-like discolorations, or, if the leakage is from a small blood-vessel, small blood spots known as petechiae. It also leads to a degeneration of bones and teeth. Spontaneous fractures frequently occur in the former and the latter become loose and fall from the gums. It is due to lack of vitamin C, a vitamin which normally occurs in fresh fruit and vegetables.

Rickets was originally thought to be due to lack of fat in the diet; actually it is due to the lack of one of the principal vitamins found in animal fats, that is, vitamin D.

It was also so common in this country at one time as to be called the "English disease". Children affected with it have distorted and twisted bones. Their legs, in particular, are usually badly bent. Sunlight can act as a preventative of rickets by causing the formation of vitamin D from a substance already in the skin.

These three diseases in their severe forms are rarely met with in England at the present time. They are final results of almost complete deprivation of the vitamins which prevent them. What is very likely, however, is that an enormous amount of general ill health, intercurrent infection, epidemics of the influenza type, poor development, and industrial inefficiency, are due to small deficiencies of vitamins extending over a long period of time. As may be seen from Sir John Orr's classic work, *Food, Health and Income*, at least 10 per cent (probably more) of the people in this country receive an income with which it is difficult to purchase enough calories, let alone adequate quantities of the vitamin-containing foods. It has been said very truly by Sherman that "half the struggle of life is the struggle for food". The millions of people living on or just below the borderline of minimum nutrition constitute a disgrace in peace time and an especial danger during a war. Any further fall in the standard of living for these people will make them an even better breeding ground for disease epidemics than they are at present. During the last war there was a considerable increase in the death-rate from tuberculosis, and at the end of the war there was a terrible influenza pandemic which swept over most of the world. Most nutritionists now agree that this was primarily due to the lowered nutritional state forced upon the people during the four years of the war. If such an

epidemic strikes England while this war is in progress
it will kill more people than a large number of air
raids.

At all costs the nutritional level must be maintained
or raised. If it is impossible for wages to follow prices,
some national scheme should be adopted for making
available supplies of vitamin-containing foods. This is
much more important even than air raid precautions,
and would probably save many more lives. It is also
vital that some special action should be taken to ensure
that the children of the nation do not suffer nutritionally
as a result of the war. The system of child allowances
which has been advocated by the Family Endowment
Society may well prove to be a partial solution of this
problem.

From these general remarks on vitamins we now pass
to a consideration of their separate significance.

VITAMIN A

Vitamin A is found in fish liver oils, e.g. cod and
halibut liver oils, in beef liver, butter, milk and eggs.
It is a colourless substance, but plants frequently contain
a yellow pigment called carotene which can be turned
into vitamin A by the human body. This yellow pigment
is the substance which gives the colour to carrots,
pumpkins, yellow peaches, and so on. Carotene also
occurs in the leaves of green vegetables, but its yellow
colour is masked by the presence of the green chloro-
phyll. It is of interest that the yellow colour of canaries
is due to the presence of a carotenoid pigment in their
feathers, and some workers claim to have deepened the
yellow colour of canaries by feeding them with foods rich
in carotene. When carotene is taken into the body, part of

it is stored in the liver, where it is believed the change into vitamin A takes place. Some of the carotene may become dissolved in the fat present in the body and cause it to take on a yellow appearance.

Vitamin A was once thought to be particularly concerned in the resistance of the body to infections, but the present view is that, while its deficiency in the diet will render the body more likely to infection, once the optimum ration is given extra amounts will not increase the resistance any further. The vitamin appears to maintain the mucous membranes of nose, mouth and respiratory organs moist and in good condition. It is necessary also for a healthy skin and eyes. Deficiency of vitamin A in the diet results in what is known as keratinisation of the mucous membranes, skin and the cornea of the eye. In India, a condition known as "toad skin" has been found to occur amongst certain people suffering from a deficiency of the vitamin. The skin in this condition becomes covered with horny outgrowths. A deficiency disease which occurred in some countries, particularly in children, at the end of the last war, was keratinisation of the outer skin of the eye—the cornea— accompanied by corneal ulcers. If not rapidly treated these ulcers result in permanent blindness.

The hardening of the cornea of the eye is known by the name xerophthalmia. This disease was known at the time of Hippocrates and there was also a cure for it. An ointment was compounded out of the juices of roasted liver and was applied to the eyes.* Sufferers were also recommended to eat liver. Both these treatments we now know would have been beneficial. Brazil, the U.S.S.R. and Denmark have all had epidemics of both

* Recent research indicates that ointments containing vitamin A are valuable in treating a number of eye diseases.

xerophthalmia and night blindness within recent years. They are still common in India and China. In fact Dr L. J. Harris of Cambridge says of xerophthalmia: "To this day in India it is still the chief cause of preventable blindness in children." Words fail to express one's emotions at the realisation that a disease, the cure of which was known over two thousand years ago, should persist in the world to-day.

The nervous system, the gums and the teeth are also affected by deficiency of this vitamin. Chronic deficiencies of vitamin A are believed to play a part in the development of kidney and bladder stones, and slight deficiencies may express themselves as night blindness. This latter condition was known to the Egyptians and Greeks before the birth of Christ; and to-day members of certain communities living upon restricted diets suffer from the disease. There is evidence that night blindness is common among many sections of the community in this country. It is a condition in which sufferers find it difficult if not impossible to see in a dim light, particularly if they have been for some time previously exposed to a bright light. It is quite possible that one of the factors involved in the alarming increase of road accidents since the institution of the blackout is vitamin A deficiency. It is certain that if the nutritional level of the people falls appreciably there will be an even greater increase in the number of these "black-out deaths". They may even reach such a degree that it would be wiser to dispense with the black-out and run the risk of air raids. The more obvious and useful thing to do, of course, is to prevent the nutritional level falling.

Vitamin A is present in the retina or "seeing part" of the eye, and recently it has been suggested that its precursor, carotene, plays an important part in colour vision.

VITAMIN B

Vitamin B includes the well-known vitamin B_1, which is the anti-beri-beri factor, and a number of others. Of these, vitamin B_2 has now been separated into lactoflavin and nicotinic acid (the pellagra-preventing factor). The present position of the group of B vitamins known as the B complex is as follows:

Vitamin B_1. Aneurin or thiamin.
Vitamin B_2. Lactoflavin or riboflavin.
Vitamin B_3. No special name.
Vitamin B_4. No special name.
Vitamin B_5. No special name.
Vitamin B_6. No special name.
The P.P. factor (pellagra-preventing factor),
Nicotinic acid.

The relationships of many of these subdivisions of the vitamin B complex to human health and disease are not yet worked out. We shall have space in this book to deal only with vitamin B_1 and the pellagra-preventing factor.

Vitamin B_1. This vitamin is found in yeast, yeast extracts and the husks and germ of cereal grains. It is contained in relatively large amounts in wholemeal and germ bread and in very small amounts in white bread. This vitamin was successfully synthesised in America in July 1936 by Dr R. R. Williams, as a sequel to some twenty years' research.

One of the first symptoms of deficiency of this vitamin is loss of appetite. Sir Robert McCarrison, in the course of his pioneer investigations into nutrition in India, showed that in vitamin B_1 deficiency there is a degeneration of Auerbach's nerve plexus in the alimentary tract. This is probably the physiological explanation of the high incidence of constipation in this deficiency. It has now been proved that vitamin B_1 is also associated with the utilisation of carbohydrate by the body. The outstanding researches of Prof. R. A. Peters and his coworkers at Oxford have taken us a long way towards understanding the part played by the vitamin in the complicated system governing the oxidation of carbohydrates in the body.

Another early symptom of vitamin B_1 deficiency is a slowing of the heart beat, and Dr L. J. Harris of Cambridge and his fellow-workers have used this fact in a very ingenious fashion to assay the amount of vitamin B_1 in various substances, particularly in urine. The vitamin B_1 content of the urine is an index of the degree of vitamin B_1 deficiency or of sufficiency in a person. Dr Harris adsorbs the vitamin B_1 from urine on to acid clay and adds this clay to the food of rats deficient in the vitamin. The time taken for the heart-beat rate of the rats to return to normal is measured, and from this result the amount of vitamin B_1 in the urine is calculated and the degree of vitamin B_1 deficiency, if present in the person investigated, obtained.

If deficiency of this vitamin is prolonged, symptoms of beriberi develop. There is tenderness, then loss of sensation in the skin, then paralysis and muscular wasting. The disease may occur in infants: in such cases loss of voice is quite common, and the child can make only a plaintive whine which is very characteristic of the condition.

It is believed that many of the cases of neuritis, which occur among the people of this country, may be due to slight deficiencies of vitamin B_1 extending over a long period of time. That dreadful facial pain, known as trigeminal neuralgia, which is so common among women, is being very successfully treated in America with injections of vitamin B_1. In chronic alcoholics there is a very painful condition, known as polyneuritis, which has also been shown in America to be due primarily to vitamin B_1 deficiency. The constant consumption of large quantities of alcohol leads to a chronic gastritis which results in failure to absorb vitamin B_1, and this appears to be the principal cause of the neuritic pains. The vitamin has to be injected if it is to alleviate the complaint.

An important factor governing the human need of vitamin B_1, and one which is frequently overlooked by the general public, is that the more food you eat the more vitamin B_1 you need. The greater the proportion of carbohydrate in your diet the more vitamin B_1 you should take. The diets of a very large proportion of the population of this country are probably deficient in vitamin B_1; this applies both to the well-to-do and to the very poor for essentially the same reason. The poorer people tend to buy carbohydrate foods which supply plenty of cheap energy; this in fact appears to be about the only way in which many of them can obtain their minimum calorie requirements. The vast majority of these people do not consume nearly enough vitamin B_1 to enable them to make full use of these carbohydrates; the result is, in effect, a vitamin B_1 deficiency, sufficient when spread over a period of years to have a serious effect on their general health and efficiency. The well-to-do probably suffer from a certain amount of vitamin B_1 deficiency as a result of those incredible carbohydrate orgies known as afternoon teas, when vast quantities of white bread, jam, cakes and scones or muffins are consumed. Even if wholemeal bread were used instead of white there would still be much too little vitamin B_1 to utilise this mass of carbohydrate. The custom of taking a few tablets of compressed dried yeast might be instituted as a ritual at such teas.

The P.P. (pellagra-preventing) factor. The P.P. factor was originally thought to be vitamin B_2, but this term is now used to describe a substance known as lacto- or ribo-flavin. The nature of the P.P. factor was for long a mystery, but has quite recently been shown to be a well-known chemical called nicotinic acid, which is related to the nicotine of tobacco. Nicotine itself, however, is a poison, and is of no value in the prevention of pellagra.

The disease is much more common than is realised. In 1912 in Roumania, out of a population of five millions, there were estimated to be 75,000 cases of the disease. In the United States of America in 1917 there were estimated to be 170,000 persons afflicted. Even now in the U.S.A. thousands die every year from it. The principal symptoms are a skin disease and diarrhoea, and in the final stages madness. At the present moment the incidence of the disease in Egypt is probably higher than in any other country.

Persons who suffer from pellagra have usually been living upon such diets as maize and fatty* meat for some time. An American physician, Dr Goldberger, eventually showed that the disease was not due to an

* The P.P. factor is contained in the lean part of the meat; if fat meat is eaten not enough lean is obtained.

infection, as had been thought by most people, or to insufficient amounts of certain proteins, but to the absence of some factor in the diet which was called the P.P. factor. In 1937 Prof. Elvehjem in the U.S.A. was led to feed dogs suffering from black tongue (the canine equivalent of pellagra) with nicotinic acid. The animals were cured. The next step was to extract nicotinic acid from liver (liver prevents pellagra and black tongue); the nicotinic acid so obtained was found to be equally effective in treating the canine disease. The final step was to try the nicotinic acid on the sufferers from human pellagra. This was immediately successful. During the recent Spanish war pellagra began to appear among the Republicans, and American scientists sent a gift of nicotinic acid to help to combat it.

Only a few cases of pellagra have been recorded in Great Britain, but it is possible that many vague skin and digestive disturbances among the population may be due to a slight deficiency of this factor.

VITAMIN C

The disease due to lack of vitamin C, namely scurvy, was the curse of early sea voyages. Vasco da Gama in his voyage round the Cape of Good Hope in 1498 lost a large number of his men from its ravages. Scurvy also decimated the armies of the Crusaders, and it has played an important part in the destruction of many armies since then. Even during the last war some of the Indian troops in Mesopotamia suffered from the disease. That scurvy was caused by the lack of something in the diet was shown in 1747 by James Lind, who found that it was rapidly cured by orange and lemon juice. The disease itself is not common at the present day. In scurvy the permeability of the blood-vessels is affected and bruises and blood spots (petechiae) develop as a result of the oozing of blood from the vessels. In addition the gums become swollen, spongy, ulcerated and bloody. The teeth become loose and fall out, the bones fracture spontaneously and if vitamin C is not given the patient dies. A form of infantile scurvy (Barlow's disease) is known; it was formerly regarded as a form of rheumatism owing to the swollen and tender condition of the joints. Mild forms of infantile scurvy are very likely to occur in bottle-fed babies unless special precautions are taken to add vitamin C in some form. It is usual to give orange juice.

Citrous fruits (i.e. oranges, lemons and grapefruit) are the best source of this vitamin, but tomatoes, berries of various sorts, rose hips, pine needles, strawberries and pineapples are very good.

The vitamin is believed to play a part in the maintenance of the connective tissues of the body in good condition.

While scurvy, as such, has practically disappeared from Great Britain, there is considerable evidence of a widespread sub-scurvy condition. The amount of vitamin C per day necessary to prevent scurvy is about seven thousandths of a gram of the pure vitamin, but various authorities give forty or fifty thousandths of a gram* as the amount necessary to prevent sub-scurvy. Therefore there is a very large margin between minimum and optimum requirements of the vitamin, and it is possible to live between these limits and be unaware of the deficiency. The effect may be in the nature of

* Recent work suggests that one hundred thousandths of a gram is nearer the optimum amount.

especial susceptibility to colds and influenza and to other infections, as there is evidence that the vitamin plays an important part in the development of immunity to certain diseases. It is possible that frequent nose bleeding is a result of insufficient intake of this vitamin.

Vitamin C has recently been shown to exert a protective action against lead poisoning; persons associated with lead or lead products in the industrial world should be given an extra ration of vitamin C to help them combat this complaint.

In view of the evidence that vitamin C is associated more than any other vitamin with resistance to infection, the possible relationship of this factor to the causation of influenza epidemics should be borne in mind. It constitutes a strong argument against any restriction of the importation of citrous fruits. Although substitutes could be obtained it would take a considerable time to organise their distribution and to educate the public in their use; and by the time all this had been done it might be too late. Young babies in particular depend upon orange juice for their vitamin C supplements.

Vitamin C has been synthesised and is known as ascorbic acid. It has been suggested recently by the Hungarian biochemist, Prof. Szent Györgyi, that the fragility of the blood capillaries which has been attributed to deficiency of vitamin C may in reality be due to lack of a closely associated vitamin which he calls vitamin P.

VITAMIN D

The classic description of rickets was given in a treatise published in England by Dr Glisson in 1650. In 1890 an English physician, Dr Palm, considered the disease to be due to lack of sunlight. For some time there was considerable doubt as to whether the disease was caused by dietary deficiency or whether it was simply due to unhygienic conditions of living. The suggestion that rickets could be prevented by sunlight was supported by the fact that investigations of Egyptian mummies revealed no trace of the disease. In 1915 Dr Edward Mellanby* investigated the effect of diet upon the occurrence of rickets in puppies. He found that the disease could be prevented with cod liver oil. After the last war rickets was very common in Berlin and Vienna; investigations carried out in those two cities showed that, not only was sunlight effective in curing rickets, but that the component of sunlight which had this effect was the ultra-violet light. In 1924 it was found that irradiating the food with ultra-violet light was as effective as irradiating the body. Finally, it was found that both skin and food contain a substance known as ergosterol, which changes to vitamin D as a result of exposure to ultra-violet light. It is of interest that sweat has been shown to contain appreciable quantities of ergosterol, and that when a person lies for some time in the sun, as many do, for example, on bathing beaches, a considerable amount of vitamin D is formed in the sweat. If one then goes for a bathe, this valuable vitamin D is washed away. If on the other hand the sweat is not washed off for some hours, the vitamin D will be absorbed by the skin.

Birds have a little gland near the tail which is called the preen gland. If you have ever observed a bird preening its feathers you will observe that every now and then it digs its beak in near its tail. It collects oil in this way from the preen gland and spreads it over

* Now Sir Edward Mellanby.

its feathers. When the feathers get their next preening a good deal of this oil, which in the meantime has been exposed to the sunlight, gets into its mouth and is swallowed. Since ordinary preen-gland oil contains ergosterol the bird gets the major portion of its vitamin D in this way. In fact, if you cut out a bird's preen gland, it will get rickets (provided it does not get vitamin D in its diet). Similarly, the elaborate toilet of many animals, the licking of its fur by a cat, the combing of the fur by the comb-like teeth of the lower jaw of a lemur, serve not only to keep the fur in condition, but also probably help to supply the animal with vitamin D obtained from the irradiation of the natural oils of the hair. The attentions of monkeys to one another, during which they appear to be searching for fleas and eating them, but which really consist of seeking for and eating pieces of scurf, are probably also associated with this problem of obtaining sufficient vitamin D.

The chief characteristics of rickets in children are a protruding forehead, curved bones in legs and arms, spinal malformation, protruding abdomen, and de-pressed ribs giving a pigeon chest. The disease if uncorrected causes life-long disablement. Rickets also causes lowered resistance to disease, particularly of the broncho-pneumonia type; and it is usually, but not invariably, accompanied by carious (decayed) teeth. Rickets may be caused, not only by deficiency of vitamin D, but also by deficiency of either of the minerals calcium or phosphorus in the diet. These minerals constitute the major percentage of bone, and their utilisation by the body is effected with the aid of vitamin D. Hence a deficiency in any one of the three factors will cause the deficiency disease to develop.

Vitamin D is found principally in cod and other fish

liver oils, in egg yolk and in butter. It is essential that
children in this country should be given a daily supplement of cod or halibut liver oil or of one of the vitamin
concentrates now on the market. A rickets-like disease
called osteomalacia is known among adult persons.

There is present in cereals a substance known as
phytic acid which tends to precipitate the calcium of the
food in an insoluble form in the alimentary tract, so that
it cannot be absorbed by the body. The effects of the
phytic acid can be overcome by taking extra vitamin D
and larger amounts of milk, but the best thing to do is
to replace most of the cereals in the diet of a young
child with other foods, such as potatoes.

Decrease in the consumption of butter during the
war as a result of rationing will cause some reduction
in the vitamin D intake of the population,* particularly
as no regulations have yet been issued making it
essential for margarine to compare favourably with
butter from the nutritional point of view.† Margarine
would probably be more useful if both vitamins A and D
were added to it. Special steps should be taken by
those in authority to ensure sufficient stocks of cod and
other liver oils, and of vitamin concentrates, during the
war, if we are to avoid an increase in the incidence and
severity of rickets.

Vitamin E

The necessity of a substance in the diet essential for
the reproduction of rats was shown by Prof. Herbert
Evans in the U.S.A. in 1922. This substance was called
vitamin E. Recently, as a result of the work of Emerson,

* Prof. J. C. Drummond has recently pointed out in *Nature* that
butter and margarine are not very important as sources of vitamins A
and D in our ordinary diets, unless we eat a lot of them.

† But see footnote on p. 51.

Evans, Drummond, and others, this vitamin has been isolated in the pure condition and has been given the name alpha-tocopherol. Its absence from the diet of the male rat has been shown by Dr Karl Mason of the U.S.A. and others to result in a profound degeneration of the testes. In a pregnant female rat suffering from vitamin E deficiency the young develop in the uterus to a certain stage and are then absorbed. Just how far these results are applicable to human nutrition has not been definitely decided,* but there is considerable evidence that a number of the cases of habitual abortion occurring in this country are due to deficiency of vitamin E in the diet.

Vitamin E is found in largest amounts in the germs of cereals and in lettuce. Wheat germ oil provides a dietary supplement very rich in the vitamin.

Vitamin K

This vitamin was found in 1934 to be essential for the good health of chicks. In its absence there is a tendency to uncontrollable haemorrhage (bleeding). We do not know yet whether this vitamin is essential for human nutrition, but it has been used with success in treating haemorrhages associated with certain human diseases. It has recently been isolated and synthesised by Prof. Doisy and his collaborators in St Louis, U.S.A.; its chemical name is naphthoquinone. Vitamin K is found in greatest amounts in fish meals, after they have undergone some putrefaction, the liver fat of pigs, in spinach, and in alfalfa.

* Recent research in England has shown that vitamin E may play an important part in treating and possibly preventing certain neuro-muscular diseases; it is thought that its deficiency in the diet may accentuate the muscular degeneration which often follows infantile paralysis.

CHAPTER IV

Minerals

According to Prof. Sherman, the American nutritionist, the following elements occur in the given proportions (by weight) in the body:

	Approx. percentage		Approx. percentage
Oxygen	65·0	Sulphur	0·25
Carbon	18·0	Sodium	0·15
Hydrogen	10·0	Chlorine	0·15
Nitrogen	3·0	Magnesium	0·05
Calcium	1·5	Iron	0·004
Phosphorus	1·0	Iodine	0·00004
Potassium	0·35		

Manganese, fluorine and silicon. Very minute amounts.
Aluminium, cobalt, nickel and copper. Traces.

Of these elements, calcium, phosphorus, potassium, sulphur, sodium, magnesium, iron, iodine, manganese, fluorine, silicon, aluminium, cobalt, nickel and copper can be classified as mineral substances. Potassium and sodium are extremely common in our foodstuffs and there is hardly likely to be any deficiency of them. Particularly is this so in the case of sodium, which we use rather lavishly in the form of common salt. Iron, copper and cobalt, and possibly manganese and nickel, are associated with the production of the red pigment of the blood. Iron is required in the largest amount because it actually forms part of the haemoglobin molecule, but the others are needed only in minute traces; they probably act as catalysts.

Sulphur is a constituent of nails, hair and skin. There does not seem to be any record of its deficiency in the human diet.

The significance of aluminium, fluorine and silicon in nutrition is obscure.

Iodine is essential for the production of the hormone of the thyroid gland, since it occurs in the hormone. It is most commonly deficient in mountainous regions, and here, as a result, goitre is a common disease. Iodine is usually contained in appreciable quantities in drinking water. Fish is a good source of iodine, particularly fish roes.

The average man excretes from his body in the urine something like 30 grams (1 oz.) of mineral salts per day. Prof. Sherman lists the functions of minerals in the body as follows:

1. As constituents of bones and teeth.

2. As essential elements of organic compounds which are the chief solid constituents of soft tissues.

3. As soluble salts (electrolytes).

About one-fifth of the body weight is made up of bones and teeth, formed principally from calcium and phosphorus. If there is a deficiency of these substances in the food the body can call on the bones to a certain extent to make good the deficiency. Both bones and teeth are frequently seriously depleted in women during pregnancy and nursing owing to the need of the baby for these minerals; this, of course, can only occur when the diet of the mother does not contain enough of them.

Calcium also plays an important part in the activity

of muscle and nerve, and in the clotting of blood. Phosphorus is associated with the breakdown of carbohydrates by the body, and it also forms an important constituent of the phospholipoid substances of the brain.

Potassium and sodium form a balance in the blood and tissues, and their variations in concentration are important in connection with "shifts" of water from one tissue to another.

Iron, as mentioned before, is an essential constituent of haemoglobin, the blood pigment responsible for carrying oxygen from the lungs to the various tissues. Copper and possibly manganese and nickel play a part in the formation of this haemoglobin, and it has recently been found that in sheep and cattle, at least, the presence of cobalt is essential for normal blood formation.

The human requirements of the various minerals are as follows:

Calcium

Adults require 0·4 to 1·0 gram of calcium (Dodds and Robertson, 1939).

Women during pregnancy and lactation need at least a gram of calcium a day and preferably more.

Children require about 1 gram of calcium per day.

Phosphorus

Adults, 1·3 grams per day.

Women during pregnancy and lactation, 1–2 grams per day.

Children, about 2 grams per day.

Iron

Men, twelve thousandths of a gram per day.

Women, fifteen thousandths of a gram per day.

Pregnant women, twenty thousandths of a gram per day.

Children, five ten thousandths to one thousandth per day.

Babies are born with a certain amount of iron stored in their livers, and while they are suckling or being fed from the bottle they are drawing upon this supply. At about the fifth or sixth month their iron supplies are considerably diminished and it is advisable to supply them with this mineral in the form of vegetable purées, lightly boiled egg yolk, etc., before this.

The best sources of calcium and phosphorus are milk, cheese, egg yolk and almonds. Iron is found in relatively large quantities in dried beans and peas, egg yolk, meat, dried prunes, spinach and watercress. Nuts, cereals and dried fruits are good sources of copper.

We need calcium, phosphorus and iron in relatively large quantities in our diets, and with them we usually absorb sufficient of the other minerals to supply the traces required; consequently, if we make sure that we obtain adequate supplies of the three principal minerals from natural sources, then we need not worry about the other minerals.

CHAPTER V

Nutrition in War Time

Foods in general may be divided into two main groups, "protective foods" and "non-protective foods". The former are rather rich in minerals and vitamins but are, with a few exceptions, rather low in calories; the latter usually supply plenty of cheap energy but are deficient in the minerals and vitamins. To a certain extent the two groups of foods are complementary, but the more you derive your food from the protective group the more you will approximate to a proper balance between calories, minerals and vitamins. Although carbohydrate-containing foods are frequently deficient in minerals and vitamins they may be quite valuable in the diet. For example, many of the protective foods are rich in fat; and fat can only be burnt properly in the body if a certain amount of carbohydrate is present. If there is not sufficient, the fat will be incompletely burnt in the body and a condition known as ketosis will result.

The non-protective group of foodstuffs includes a number of highly refined products, such as white bread, jam, cakes and sugar, which require large quantities of protective foods to balance them. If the former constitute a large proportion of the diet, then vitamin supplements, such as dried yeast, wheat germ, etc., must be taken to compensate for them.

The protective foods include dairy products (milk, butter, eggs, cheese and cream), fresh fruit, vegetables, fresh meat or fish and wholemeal or germ bread. The non-protective foods are represented by white bread, sugar, macaroni, spaghetti, refined cereals, polished rice, most cakes, jams and biscuits.

A word about honey may be of interest. Honey is a very good source of easily digestible carbohydrate and it contains some minerals as well. It is, however, almost completely deficient in vitamins. Its superior mineral content makes it a much better sweetening agent than refined sugar.

During the last war Miss Gillett, who was associated with the United States of America Food Administration, recommended that the money to be spent on food should be divided into fifths and that the various fifths should be spent as follows:

1st fifth.   Vegetables and fruit.

2nd fifth. Milk and cheese.

3rd fifth. Meats, fish and eggs.

4th fifth. Bread and cereals.

5th fifth. Fats, sugar, other groceries and extras.

This is, of course, the desirable distribution, but in many cases the income is so small that if the money is spent in this way there are not enough calories to prevent hunger. This system should, however, be borne in mind and an attempt made to follow it as closely as possible.

The ability of a food to satiate the appetite, "satiety value" as it is called, is no indication of its nutritional value: such diverse substances as milk and refined sugar have a high satiety value, while bread has a low value. The following are recognised as rapidly satisfying hunger: sugar, butter, potatoes, milk, meat, eggs, fatty fish, and fatty foods in general. Those foods which have a low satiety value are: bread, all green vegetables, and non-fatty fish.

Prof. J. C. Drummond of London University has emphasised that "there is no security in placing trust in the belief that what is vaguely termed a mixed diet will protect against deficiencies in one or another foodstuff. It is unfortunately a view that is widely held and frequently expressed. There are many examples of relying upon this belief." The diet therefore must be planned to include those foods known to be rich in the various essentials and the scheme proposed by Miss Gillett is earnestly recommended to all those who can afford it.

The importance of an allotment or a garden wherein to grow vegetables, so important from the point of view of vitamins and minerals, cannot be over-emphasised at a time like the present.

At this point it may be well to add a note on the cooking of foodstuffs. Overcooking and twice cooking should be avoided as far as possible. Take as an example the effect of heat on cabbage. In the raw state 1 oz. of cabbage per day will protect an adult from scurvy. If the cabbage is boiled for 20 minutes 4 oz. will be required to protect from scurvy, if it is boiled for 1 hour 10 oz. will be required. If it is boiled with soda for only a short time it will be valueless as a protective against scurvy. Jam, which is made with long heating and stirring of fruits, contains no vitamin C. Marmalade is made from oranges, a fruit which normally contains a large amount of this vitamin, but as a result of the long cooking which it receives its vitamin C content is negligible. Cooking meat or potatoes or other vegetables twice usually destroys all traces of even the more heat-stable vitamins such as vitamin B_1. The water

in which vegetables have been boiled should not be
wasted because it contains many of the mineral salts of
the vegetable; it should be made into a soup, preferably
a thick one with peas, beans, lentils or potatoes;
alternatively, it could be made into a gravy and used
to pour over meat.*

At a time such as this, when special care must be
taken to avoid food deficiencies, it is doubly important
to use wholemeal flour instead of white flour as far as
possible in cooking. Quite pleasant cakes can be made
from wholemeal flour, and bran (a potent source of
vitamin B_1) can be made into excellent biscuits. Of
course it may not always be possible to obtain wholemeal
flour, since in time of war so much of the bran part of
grain is required as animal food.

If you have the opportunity of obtaining skim milk
remember that it is still a valuable food, even though
most of its fat and fat-soluble vitamins have been
removed. It still contains most of its first-class protein
and practically all its valuable minerals. Do not replace
your whole milk with skim milk just because you can
get it more cheaply, but buy some in addition to your
regular supply of whole milk, particularly if you have
young children in the house.

The chief factor which is disturbing the minds of
many people in connection with rationing is the cutting
down of supplies of butter. Although butter is a desirable
addition to our diet it can be supplanted by other things.
Butter contains principally fat and fat-soluble vitamins A
and D. In the place of butter we are being offered
margarine. Margarine is made usually from whale oil,

* It is of interest also that frying food in fat considerably increases the calorific value.

which has been hardened by a process known as hydro-genation. Some liquid vegetable oil is sometimes mixed with it. It is therefore composed almost entirely of fat, but without the essential vitamins A and D. Some of the better quality margarines contain these vitamins, but we have no idea how much. It should be made illegal, first, to sell any margarine without vitamins A and D; and, secondly, to sell it without the amount of vitamins A and D (in international units) which have been added per pound being shown on the wrapper. At present any manufacturer is permitted to add one drop of vitamin A and D concentrate to any quantity of margarine he likes and call the result vitaminised. As a result, even though we can purchase vitaminised margarine, we have no idea whether the vitamin content approximates to the average vitamin content of butter or not. Even butter varies considerably in its vitamin content, and the standardisation of the vitamin content of butter by the adding of concentrates to those samples shown by analysis to be low might well be undertaken as soon as the present war has concluded, or even before. In any case it would be a valuable aid to health if everyone would supplement the vitamin A and D content of his diet by taking good quality cod liver oil or halibut liver oil or concentrates.* It would be a very valuable war measure if the government would take control of all stocks of vitamin A and D concentrates, cod liver oil,

* As this book goes to press it is announced in the press by the Ministry of Food that from 5 February 1940 all margarine other than that sold for manufacturing purposes will incorporate the vitamins A and D in the same quantity as they are found in butter. There is no official announcement, however, that any check is to be kept on the vitamin content, and in any case butter can hardly be regarded as an accurate vitamin standard.

and so on, and ensure that everyone in this country received adequate supplies of these foods irrespective of his capacity to pay for them.

It is of interest that the herring, which is a fatty fish, is a valuable source of vitamin D. One hundred grams (roughly ¼ lb.) of herring contains 600 to 1000 units of vitamin D.

The daily vitamin D requirement of an adult is somewhere in the region of 1000 international units, that of children from 1500 to 2000 international units. Tinned salmon also contains vitamin D, but the amount of vitamin present depends upon the fat content of the salmon, and the amount of variation of fat in this fish is considerable. For example, tinned Sockeye salmon commonly sold in Australia contains about 5 per cent of fat, while American analyses of tinned Sockeye salmon show 11 per cent of fat. In England it varies between these two figures. The fat content, and therefore also the vitamin D content, of herrings is also subject to seasonal variation.

Cod liver oil supplies both vitamins A and D. Vitamin D can, however, be made synthetically on a very large commercial scale. It would not be a very costly matter to organise the industry so that every child in the country was assured of an adequate supply of vitamin D. According to A. L. Bacharach, the annual cost of producing enough vitamin D per year to protect every child in this country from rickets would be only £100,000: the cost of two or three bombing aeroplanes.

VITAMIN OVERDOSAGE

It may be as well to mention here the subject of vitamin overdosage. If you can afford to derive all your vitamins from protective foods instead of supplementing a bad diet with vitamin concentrates you can dismiss any possibility of this. In any case it is impossible to give an overdose of the water-soluble vitamins B and C, because any excess is so rapidly excreted in the urine. Vitamins A and D, however, being fat soluble, may be given in overdose. When one realises, however, that the overdose for these vitamins is in the case of vitamin D something like twenty or thirty times, and in the case of vitamin A one thousand times that required for a normal dose, it is obvious that such a condition is not likely to occur unless there is a gross misuse of vitamin concentrates. If you depend upon such vitamin concentrates for providing your children with adequate vitamins A and D you should give only the dose recommended by your physician.

FOOD SUPPLIES

We may next examine the origin of our chief sources of food, because it is imported foods that we are most likely to have difficulty in obtaining in adequate quantities. Our milk is of course, as a matter of necessity, entirely home produced; but some of the feeding stuffs used for cows are not, so that if there is any decrease in the supplies of oil cake, by-products of imported wheat, etc., there may well be some difficulty in supplying enough milk for home consumption. This is probably one of the most vital of the protective food problems, because milk is a protective food *par excellence*, and any considerable reduction in our supplies is likely to cause widespread ill-health, particularly amongst children.

Most of our potatoes are home produced, and plans

have already been made to increase this production considerably. This is most important, because potatoes contain good quality protein and have a high percentage of carbohydrate (making them a good source of energy). They also contain a small but useful amount of vitamin B_1. A time may come, although ·I earnestly hope that it will not, when imports of fresh fruits will be restricted; in that case, potatoes will be our chief source of vitamin C, and if there is any failure of the crop there may be widespread outbreaks of scurvy. We cannot obtain our optimum intake of vitamin C from potatoes alone; we need fresh fruits and preferably citrous fruits. There are of course supplies of synthetic vitamin C available, but the members of the lower income groups may not be able to purchase sufficient for their needs.

We import about 50 per cent of our fresh meat, so that, even if no meat supplies at all come through, which is unlikely, we shall only be reduced to half rations in this commodity. This will not be serious, because we can make up our protein supply from foods such as cheese, bacon (which also supplies valuable fat), fish and eggs; all of these contain valuable vitamins and minerals as well as first-class protein. If we have difficulty in getting our protein from these sources we can still obtain it in its second-class form from vegetables such as peas, beans, lentils, nuts and potatoes.

Britain imports something like 90 per cent of her fats and oils, and it is these which seem most likely to be decreased in amount. Animal fats usually contain vitamins A and D, but in addition they are valuable reserve stores of energy. Mineral oils, such as paraffin oil, are of course no use as food. As far as we know, there is no special reason why we should have animal fats instead of vegetable fats, apart from the vitamin A and D content; if therefore you have difficulty in obtaining animal fats do not be unduly disturbed provided you can obtain a little cod or halibut liver oil to supply you with the fat-soluble vitamins.

There is a number of novel substances which could be used for food. The gypsies are reputed to relish hedgehogs baked in clay, and badgers are believed to be quite edible. We must not forget the rabbit, although it can hardly be called a novelty. Rabbit flesh is rather poor in fat, and has therefore less calorific value than the flesh of a bullock; but it is a valuable source of first-class protein.

According to Dr E. G. Boulenger the Canadian grey squirrel, the musk rat and the coypu, which are farmed in this country for their furs, make good eating, and there is no reason why such flesh should not be eaten more widely if it is obtainable. Dr Boulenger also points out that grass snakes are eaten in France under the name of "hedge eels", that tinned rattlesnake and fresh-water tortoise can also be used as food. He mentions that python steak is a delicacy throughout Africa, but Prof. J. B. S. Haldane, who boiled for himself a foot of Zoo python when it was killed at the outbreak of war, and had it for his lunch with potato chips, regarded it as being indistinguishable from boiled string. Dr Boulenger also regards the slipper limpet, a pest of the oyster beds, as a potentially valuable source of food, and he points out that octopus and squid may be purchased in many Soho restaurants. There is no doubt that our marine biologists and naturalists could advise us concerning many animals and plants which are not normally used for food but which might be pressed into service if the need arose.

One other point to which attention might be drawn is the recently developed large-scale water-culture method for growing vegetables which is now being used in America. At Wake Island, one of the stopping places for the China Clipper on its way from California to Hong Kong, fresh fruits and vegetables for use on the plane are all grown by this method. There is no reason why water culture should not be used here. Those who have no garden plot or allotment could easily supplement their vegetable supply even if it is only by sprouting mustard and cress.

Even in peace time attention to diet is of vital importance. As Prof. J. C. Drummond and Miss Wilbraham put it in their recently published classic, *The Englishman's Food*: "The experts are convinced that the national physique and general health would be greatly benefited by a large increase in the consumption of dairy produce, fruit and vegetables." This is shown by the increased height of children over their parents when they are given a better diet. For example, we think of the Japanese as being characteristically a small race, yet Japanese children born in California in the U.S.A. show an increase in height over that of their parents, often by as much as 6 inches. The Chinese are also regarded as short people; yet when they are fortunate enough to live in Hawaii, where they obtain much better food than at home, they show an increase of some inches over Chinese of a similar age in their native province in China. It is also frequently noticed that children born in America of European parents are taller and of superior physique to their parents. The same conclusion has been drawn from a number of English experiments on children. Those receiving extra milk increased significantly in height and improved in physique and general health over a similar group which did not have the milk. We have no space here, unfortunately, to go into the details of these experiments, but we might well bear in mind that Sir Francis Freemantle has estimated the cost of ill-health to this nation at three hundred million pounds a year, or nearly one million pounds a day. A great deal of this ill-health can probably be traced to improper nutrition, and when one realises that we have to foot such a bill as this in peace time it should make us doubly careful to prevent any further ill-health due to faulty nutrition in war time. It is a national service to keep oneself healthy, and attention to the diet will play a major part in maintaining normal health.

In view of all these facts one is inclined to doubt the wisdom of the leading article which appeared in the *Economist* for 11 November 1939. It stated: "The official attitude on food rationing seems to show no conception of the fact that we cannot exert our maximum effort without *deliberately restricting the consumption of food*,* whether the food is physically available or not."

There is certainly a number of luxury foods which could be restricted, but from a physiological point of view, any extensive restriction of food would be very undesirable as a home-front measure, for it would reduce rather than increase our war effort.

SUMMARY OF THE CONSTITUENTS OF FOOD

Food contains:

1. *Proteins, fats and carbohydrates*, which supply energy by means of which we keep warm and perform

* Author's italics.

work of various sorts. In addition, proteins are body building.

Proteins are of two sorts: first-class protein derived from animals and second-class protein derived from plants. First-class proteins are more economically converted into flesh.

Fats are of two sorts, animal and vegetable. They both have the same calorific value, but the animal fats usually contain vitamins A and D.

Carbohydrates include starches and sugars; they are usually the cheapest forms of energy.

2. *Vitamins.* Vitamins A and D are contained in animal fats. Vitamin A prevents eye troubles and protects against respiratory tract infections. Vitamin D prevents rickets.

Vitamin B_1 and the other "B vitamins" protect against beriberi and pellagra and certain nervous conditions. They are contained in yeast, wholemeal and germ bread, marmite, and green vegetables. Vitamin B_1 is not very much affected by cooking, unless it is cooked with soda.

Vitamin C protects against scurvy and infections. It is found in oranges, lemons and to a lesser extent in other fresh fruits and vegetables. It is liable to be destroyed by cooking.

Vitamin E occurs in wholemeal and germ bread and lettuce. Helps in the bearing of healthy children.

3. *Minerals.* Calcium and phosphorus are needed for bones and teeth, and proper functioning of muscle and nerve. Found principally, and in the best ratio to one another, in milk and cheese.

Iron is used in making blood and is obtained from meat, curry powder, egg yolk and liver.

CHAPTER VI

The Nature and Composition of Various Foodstuffs

The following list of foodstuffs and figures for protein, fat and carbohydrates is taken mainly from Technical Communication No. 10 of the Imperial Bureau of Animal Nutrition (May 1938). The calorific value given for each food is calculated approximately from the figure given in the above communication.

This list of foodstuffs does not of course include every item of food used in this country, but it does include most of the common foods.

In the list which follows an attempt will be made to tell you where the foodstuff comes from if its origin is not clear, and whether it is good, bad or indifferent as far as its calories, its protein, fat and carbohydrate, and its minerals and vitamins are concerned. At the end of this first list (List A) you will find a second list (List B) composed of the following headings: calories, first-class protein, second-class protein, carbohydrate, animal fat, vegetable fat, various minerals and vitamins. Under each of these headings will be found a list of foods which contain them in the greatest amount.

If you are unable to obtain any particular food, look it up in list A, see what it contains, then look up those things in list B and see what other foods will give them to you.

Take, for example, "butter". If you look up list A you will find that it is very rich in calories and that most of them come from fat, that it has a small amount of minerals and that it usually supplies a good amount of vitamins A and D, but that it contains negligible quantities of the other vitamins. Therefore in the place of butter, nutritionally speaking, you need a food containing a large amount of fat and a source of vitamins A and D. If you now look up fat in list B you will find that you can get animal fat from dripping and animal or vegetable fat* from margarine. If you look up vitamin A and vitamin D you will find that both of them can be obtained from halibut or cod liver oil. Therefore if you eat margarine or dripping and take either of the two fish oils mentioned you will be missing nothing from your food that you would have obtained from butter.†

* According to whether it is made from whale oil or vegetable oils.

† See footnote page 51.

LIST A

KEY TO LIST

Calories = Calories per pound of the edible portion of the food.
The term vitamin B group refers to the vitamin B complex.
The various constituents of each food are appraised as follows:

Very good	Poor
Good	Very poor
Fairly good	Negligible
Fair	

This appraisement list is, of course, approximate, but it will serve as a guide to the value of the different foodstuffs from the point of view of most constituents. Anything that is labelled "fair" or better is a valuable source of the particular constituent referred to. Anything below "fair" is not valuable.

Although many of the vegetables are disappointing in many constituents it must be remembered that they are also sources of those minerals (such as nickel, manganese, etc.) which we need only in traces, but which are essential for the proper functioning of the body. Also fruits and vegetables leave an alkaline ash in the body which balances the acid ash of starchy foods, sugars, meat and eggs.

It will probably be noticed that dried peas, beans and lentils are good in almost every constituent, but it should be remembered that we eat relatively small quantities of them at a time, so that in their case the appraisement may be a little misleading.

MILK AND MILK PRODUCTS

Raw Milk

376 calories (per pint): poor

Proteins	3·4 p.c. poor
Carbohydrates	4·8 p.c. poor
Fats	3·5 p.c. poor

Iron: very poor
Phosphorus and calcium: very good
Vitamin A: fairly good (variable)
Vitamin B group: fair
Vitamin C: poor
Vitamin D: usually good (variable)
Vitamin E: negligible

Skim Milk. Milk from which cream has been removed.

204 calories (per pint): poor

Proteins   3·4 p.c. poor
Carbohydrates   3·8 p.c. poor
Fats   0·5 p.c. very poor

Minerals: same as in raw milk
Vitamins A and D removed with the fat
Other vitamins: the same as raw milk

Condensed Milk. Whole milk which has been evaporated down in a partial vacuum and to which cane sugar has been added.

1558 calories: good

Proteins   9·3 p.c. fair
Carbohydrates   52·8 p.c. very good
Fats   10·6 p.c. fairly good

Iron: very poor
Phosphorus and calcium: very good
Vitamins A and D: fairly good
Vitamin B group: poor
Other vitamins: negligible

Skimmed Condensed Milk. In this food most of the fat has been removed before evaporation.

1287 calories: fairly good

Proteins   10·4 p.c. fairly good
Carbohydrates   57·4 p.c. very good
Fats   0·7 p.c. very poor

Iron: poor
Phosphorus and calcium: very good
Vitamins A and D: poor
Other vitamins: same as in condensed whole
   milk

Dried Milk. There are various ways of making dried milk powders; in one method the milk is sprayed on heated rollers and the water evaporates, leaving the powder, which is scraped off and put into tins. In skimmed milk the fat is removed first.

Dried Whole Milk   **2125 calories: very good**

Proteins   24·5 p.c. very good
Carbohydrates 35·1 p.c. very good
Fats   24·2 p.c. very good

Iron: fairly good
Phosphorus and calcium: very good
Vitamins A and D: usually good
Vitamin B group: fairly good
Vitamin C: negligible
Vitamin E: probably negligible

Dried Skim Milk   **1595 calories: good**

Proteins   32·8 p.c. very good
Carbohydrates 49·8 p.c. very good
Fats   1·4 p.c. very poor

Minerals: same as in dried whole milk
Vitamins A and D: very poor
Other vitamins: same as in dried whole milk

Cream (fresh)   **1792 calories: good**

Proteins   2·3 p.c. very poor
Carbohydrates 7·3 p.c. fair
Fats   38·0 p.c. very good

Minerals: poor.
Vitamins A and D: good
Other vitamins: very poor

Cream (tinned)   **1186 calories: fairly good**

Proteins   2·6 p.c. very poor
Carbohydrates 7·3 p.c. fair
Fats   23·5 p.c. very good

Minerals and vitamins: as for fresh cream

Ice Cream. Usually made from cream and milk with added cane sugar and gelatine

1441 calories: fairly good

Proteins   4·3 p.c. poor
Carbohydrates 28·0 p.c. very good
Fats   20·0 p.c. very good

Ice cream (*cont.*)   *Iron: poor*
Phosphorus and calcium: fair
Vitamins A and D: fairly good (*variable*)
Vitamin B group: poor
Other vitamins: negligible

Butter   **3497 calories: very good**

Proteins   0·2 p.c. very poor
Carbohydrates None
Fats   83·0 p.c. very good

Minerals: very poor
Vitamins A and D: usually very good
(variable)
Other vitamins: negligible

Margarine. Is made usually from whale oil by "hardening".
It consists almost entirely of water and fat. The calorific
value of margarine is therefore very high (3800 to 4200 per
pound). It has no minerals. Normally it contains no
vitamins unless vitamins A and D are added by the
manufacturers.

Cheese. Cheese is made by allowing milk to clot either with
rennet or naturally by souring. Dutch cheeses are obtained
by skimming the fat off the milk before it is clotted, and
therefore they contain very much less fat than ordinary
cheeses. The clot is pressed to remove the whey and it is
stored in order to ripen. This process is brought about by
the activity of different kinds of bacteria. Various bacteria
produce different flavoured cheeses.

1862 calories: good

Proteins   28·0 p.c. very good
Carbohydrates 2·0 p.c. very poor
Fats   31·0 p.c. very good

Iron: poor
Phosphorus and calcium: very good
Vitamins A and D: good

Cheese made from skimmed milk (Dutch Cheese)

879 calories: fair

Proteins	33·0 p.c. very good
Carbohydrates	3·0 p.c. very poor
Fats	5·0 p.c. poor

Iron: poor
Phosphorus and calcium: good
Vitamin B group: probably fair
Other vitamins: doubtful

Lactic Cheese

1780 calories: good

Proteins	22·8 p.c. very good
Carbohydrates	— negligible
Fats	32·2 p.c. very good

Iron: poor
Phosphorus and calcium: good
Vitamins A and D: fairly good
Other vitamins: doubtful

EGGS

Hen Eggs

81 calories (in one egg): fair

Proteins	12·4 p.c. fairly good
Carbohydrates	0·7 p.c. very poor
Fats	11·5 p.c. fairly good

Iron, phosphorus and calcium: good
Vitamins A, B group and D: good
Vitamin C: negligible
Vitamin E: poor

Duck Eggs

111 calories (in one egg): fair

Proteins	12·2 p.c. fairly good
Carbohydrates	0·7 p.c. very poor
Fats	15·6 p.c. fairly good

Minerals and vitamins: as for hen eggs

MEATS
Beef

Silverside

1314 calories: fairly good

| Proteins | 16·1 p.c. good |
| Carbohydrates | — negligible |

Silverside (*cont.*)   Fats   24·1 p.c. very good
Iron: good
Phosphorus and calcium: poor
Vitamins A and D: fair. Dependent upon
    amount of fat
Vitamin B group: fair
Vitamin C: poor even in fresh meat
Vitamin E: doubtful

Aitch Bone   **1658 calories: good**
Proteins   12·7 p.c. fairly good
Carbohydrates   —   negligible
Fats   33·8 p.c. very good
Minerals and vitamins: as for silverside

Brisket   **2107 calories: very good**
Proteins   10·1 p.c. fairly good
Carbohydrates   —   negligible
Fats   45·6 p.c. very good
Minerals and vitamins: as for silverside

Plate   **2306 calories: very good**
Proteins   10·7 p.c. fairly good
Carbohydrates   —   negligible
Fats   50·0 p.c. very good
Minerals and vitamins: as for silverside

Chuck steak   **1318 calories: fairly good**
Proteins   15·8 p.c. good
Carbohydrates   —   negligible
Fats   24·4 p.c. very good
Minerals and vitamins: as for silverside

Shin   **802 calories: fair**
Proteins   18·5 p.c. good
Carbohydrates   —   negligible
Fats   10·9 p.c. fair
Minerals and vitamins: as for silverside

Topside   **1314 calories: fairly good**

Proteins   16·1 p.c. good
Carbohydrates   —   negligible
Fats   24·1 p.c. very good
Minerals and vitamins: as for silverside

Sirloin   **1726 calories: good**

Proteins   12·9 p.c. fairly good
Carbohydrates   —   negligible
Fats   35·3 p.c. very good
Minerals and vitamins: as for silverside

Rump Steak   **1925 calories: good**

Proteins   11·9 p.c. fairly good
Carbohydrates   —   negligible
Fats   40·5 p.c. very good
Minerals and vitamins: as for silverside

Kidney   **444 calories: poor**

Proteins   18·1 p.c. good
Carbohydrates   —   negligible
Fats   2·6 p.c. very poor

Iron: very good
Phosphorus and calcium: fair
All vitamins except vitamin E: probably fair

Liver   **584 calories: fair**

Proteins   19·9 p.c. good
Carbohydrates (glycogen) 4·4 p.c. poor
Fats   3·2 p.c. poor

Iron: very good
Phosphorus and calcium: fair
Vitamins A, B group and D: very good
Vitamin C: fairly good
Vitamin E: probably present; amount doubtful

Beef Sausages. According to Sir Robert Hutchison and Prof. V. H. Mottram: "Sausages are preparations of very uncertain composition. It has been remarked of them with some truth that they are like life; for you never know what is in them till you have been through them." They are usually composed chiefly of bread and meat with some seasoning and colouring matter added.

1237 calories: fairly good

Proteins	11·2 p.c. fairly good
Carbohydrates	15·2 p.c. good
Fats	17·7 p.c. good

Iron: good
Phosphorus and calcium: poor
Vitamin content: doubtful
Possibly some small quantities of vitamins A, B and D

Tongue

1101 calories: fairly good

Proteins	15·6 p.c. good
Carbohydrates	— negligible
Fats	19·2 p.c. good

Minerals and vitamins: probably about the same as silverside

Tripe. Consists of the stomach and intestines of the ox after they have been washed and boiled.

267 calories: poor

Proteins	11·7 p.c. fairly good
Carbohydrates	— negligible
Fats	1·2 p.c. very poor

Minerals: poor
Vitamins: probably negligible

Sweetbread. This term is used to include both the thymus gland which is found in the chest just below the neck, and the pancreas, which is associated with the stomach and intestines. This latter organ besides secreting enzymes which help in the digestion of food secretes insulin, a hormone which controls the amount of sugar in the blood.

1078 calories: fairly good

Proteins   21·8 p.c. very good
Carbohydrates   —   negligible
Fats   16·0 p.c. good

Iron and phosphorus: good
Calcium: fair
Vitamin content: uncertain
Vitamins A, B group and D: probably fair
Vitamins C and E: poor

Heart

1178 calories: fairly good

Proteins   16·0 p.c. good
Carbohydrates   1·0 p.c. very poor
Fats   20·4 p.c. very good

Iron: very good
Phosphorus: fairly good
Calcium: poor
Vitamin B group: fairly good
Other vitamins: poor

Veal

793 calories: fair

Proteins   19·9 p.c. good
Carbohydrates   —   negligible
Fats   10·0 p.c. fair

Iron: good
Phosphorus: poor
Calcium: good
Vitamins A, C, D and E: poor
Vitamin B group: fairly good

Mutton

Neck

2111 calories: very good

Proteins   12·3 p.c. fairly good
Carbohydrates   —   negligible
Fats   44·7 p.c. very good

Iron: good
Phosphorus: fair
Calcium: poor

Neck (*cont.*)
Vitamins A and D: fair to good
Vitamin B group: fairly good
Vitamin C: poor
Vitamin E: doubtful

Shoulder

1762 calories: good

Proteins	14·3 p.c.	fairly good
Carbohydrates	—	negligible
Fats	35·5 p.c.	very good

Minerals and vitamins: as for neck

Leg

1518 calories: good

Proteins	16·7 p.c.	good
Carbohydrates	—	negligible
Fats	28·7 p.c.	very good

Minerals and vitamins: as for neck

Chops. Calorific value of chops depends upon whether both fat and lean, or lean alone, are eaten. Usually one chop will supply about 400 calories if everything except the bone is eaten.

2704 calories: very good

Proteins	10·0 p.c.	fair
Carbohydrates	—	negligible
Fats	59·8 p.c.	very good

Minerals and vitamins: as for neck

Kidney, liver and tongue of sheep are sufficiently similar to those of the ox to make it unnecessary to use separate figures for them in this list.

Pork

Bacon

2492 calories: very good

Proteins	9·0 p.c.	fair
Carbohydrates	—	negligible
Fats	55·2 p.c.	very good

Iron, phosphorus and calcium: good
Vitamins A and D: probably poor
Other vitamins: probably negligible

Ham (raw)

2224 calories: very good

Proteins   12·4 p.c. fairly good
Carbohydrates   —   negligible
Fats   47·3 p.c. very good

Minerals and vitamins: as for bacon

Leg of Pork

951 calories: fair

Proteins   19·3 p.c. good
Carbohydrates   —   negligible
Fats   14·1 p.c. fairly good

Minerals: as for bacon
Vitamins A and D: fair
Vitamin B group: fair
Vitamin C: poor
Vitamin E: probably poor

Loin of Pork

1921 calories: good

Proteins   15·4 p.c. good
Carbohydrates   —   negligible
Fats   38·8 p.c. very good

Minerals and vitamins: as for leg of pork

Meat Extracts. Contain negligible amounts of protein, fat and carbohydrates, therefore they provide practically no energy. Most of them, however, are valuable sources of phosphorus. Vitamins are probably negligible.

Brains

490 calories: poor

Proteins   10·9 p.c. fairly good
Carbohydrates   —   negligible
Fats   6·8 p.c. fair

Iron: poor
Phosphorus: very good
Calcium: fair
Vitamins A and D: probably fair
Vitamin B group: probably fairly good
Vitamin C: fair
Vitamin E: doubtful

Prepared Meats

Corned Beef

1395 calories: fairly good

Proteins   15·6 p.c. good
Carbohydrates   —   negligible
Fats   26·2 p.c. very good

Iron: good
Phosphorus: fairly good
Calcium: fair
Vitamin content: doubtful

Ox Tongue (cooked)

1359 calories: fairly good

Proteins   19·1 p.c. good
Carbohydrates   —   negligible
Fats   23·9 p.c. very good

Iron: very good
Phosphorus: good
Calcium: fairly good
Vitamins: probably poor

Veal and Ham Paste

915 calories: fair

Proteins   16·5 p.c. good
Carbohydrates   8·3 p.c. fair
Fats   10·8 p.c. fairly good

Minerals: poor
Vitamins: doubtful

Sheep's Tongue (canned)

1463 calories: fairly good

Proteins   24·4 p.c. very good
Carbohydrates   —   negligible
Fats   24·0 p.c. very good

Iron, phosphorus and calcium: fairly good
Vitamins: probably similar to fresh tongue

Boiled Ham

1278 calories: fairly good

Proteins   22·1 p.c. very good
Carbohydrates   —   negligible
Fats   20·6 p.c. very good

Iron: fairly good
Phosphorus: good
Calcium: poor
Vitamins: probably negligible

Black Puddings. These are made from blood mixed with a filling material such as bread. Seasonings are usually added.

734 calories: fair

Proteins   20·2 p.c. very good
Carbohydrates   11·7 p.c. fairly good
Fats   3·4 p.c. poor

Iron: very good
Phosphorus: poor
Calcium: fair
Vitamins: probably some of the B group
vitamins present, but vitamin content
doubtful

Meat Pie

1871 calories: good

Proteins   8·0 p.c. fair
Carbohydrates   49·2 p.c. very good
Fats   19·2 p.c. good

Iron: fair
Phosphorus and calcium: poor
Vitamins: poor

Potted Meat

938 calories: fair

Proteins   23·6 p.c. very good
Carbohydrates   —   negligible
Fats   11·8 p.c. fairly good

Iron: fairly good
Phosphorus: good
Calcium: fair
Probably small amounts of vitamins A, B
and D present

POULTRY AND GAME

Fowl **598 calories: fair**

Proteins   24·3 p.c. very good
Carbohydrates   —   negligible
Fats   3·5 p.c. poor

Iron: fair
Phosphorus: good
Calcium: fair
Vitamins A and D: fair
Vitamin B group: probably fair
Other vitamins: negligible

One fowl weighing altogether about 3½ lb. will provide approximately 1256 calories. (60% edible.)

Duck **2333 calories: very good**

Proteins   12·1 p.c. fairly good
Carbohydrates   —   negligible
Fats   50·0 p.c. very good

Minerals: similar to fowl
Probably more vitamins A and D than in fowl
Other vitamins: probably the same

One duck weighing about 8½ lb. will provide approximately 5850 calories. (65% edible.)

Goose **1454 calories: fairly good**

Proteins   18·3 p.c. good
Carbohydrates   —   negligible
Fats   26·4 p.c. very good

Minerals and vitamins: same as duck

Turkey **743 calories: fair**

Proteins   22·5 p.c. very good
Carbohydrates   —   negligible
Fats   7·7 p.c. fair

Minerals and vitamins: similar to fowl

One turkey weighing about 8½ lb. provides approximately 3890 calories. (56% edible.)

Rabbit   **548 calories: fair**

Proteins   22·2 p.c. very good
Carbohydrates   —   negligible
Fats   3·2 p.c. poor

*Minerals and vitamins: probably similar
to fowl*

One 2 lb. rabbit supplies approximately 760 calories.

Hare   **485 calories: poor**

Proteins   21·9 p.c. very good
Carbohydrates   —   negligible
Fats   1·9 p.c. very poor

*Minerals and vitamins: probably similar
to fowl*

Cooked Fowl   **888 calories: fair**

Proteins   27·9 p.c. very good
Carbohydrates   —   negligible
Fats   8·8 p.c. fair

*Minerals and vitamins: probably similar
to fowl*

FISH

White Fish (cod,   **353 calories: poor**
haddock, ling
and whiting)   Proteins   18·1 p.c. good
                  Carbohydrates   —   negligible
                  Fats   0·4 p.c. very poor

Iron: fair
Phosphorus and calcium: good
Vitamins A and D: probably poor
Other vitamins: probably very poor

Flat Fish (flounder, **376 calories: poor**
halibut, hake,
plaice, sole and Proteins   17·6 p.c. good
turbot)   Carbohydrates   —   negligible
                  Fats   1·3 p.c. very poor

Flat Fish (*cont.*)

Minerals: similar to white fish
Vitamins A and D: fairly good
Large amounts of vitamins A and D in the
*   liver, particularly of the halibut*
Other vitamins: probably poor

Fat Fish (herring, mackerel and salmon)

911 calories: fair

Proteins   18·4 p.c. good
Carbohydrates   —   negligible
Fats   13·5 p.c. fairly good
Minerals: similar to white fish
Vitamins A and D: good
Other vitamins: poor

Eel

1305 calories: fairly good

Proteins   14·1 p.c. fairly good
Carbohydrates   —   negligible
Fats   24·7 p.c. very good
Minerals: similar to white fish
Vitamins A and D: probably very good
Other vitamins: poor

Skate

458 calories: poor

Proteins   24·3 p.c. very good
Carbohydrates   —   negligible
Fats   0·2 p.c. very poor
Minerals: similar to white fish
All vitamins probably poor

Canned Fish

Salmon

915 calories: fair

Proteins   21·8 p.c. very good
Carbohydrates   —   negligible
Fats   12·1 p.c. fairly good
Minerals: similar to white fish
Vitamins A and D: good
Other vitamins: doubtful

Sardines in Oil	**997 calories: fair**		
	Proteins	24·8 p.c.	very good
	Carbohydrates	—	negligible
	Fats	12·7 p.c.	fairly good

Minerals: similar to white fish
Vitamins A and D: probably fairly good
Other vitamins: poor

Smoked Fish

Haddock	**376 calories: poor**		
	Proteins	19·5 p.c.	good
	Carbohydrates	—	negligible
	Fats	0·3 p.c.	very poor

Minerals: same as white fish
All vitamins poor

Kipper	**974 calories: fair**		
	Proteins	18·8 p.c.	good
	Carbohydrates	—	negligible
	Fats	14·8 p.c.	fairly good

Minerals: similar to white fish
Vitamins A and D: probably fairly good
Other vitamins: poor

Dried fish

Cod or Whiting	**1454 calories: fairly good**		
	Proteins	75·0 p.c.	very good
	Carbohydrates	—	negligible
	Fats	1·5 p.c.	very poor

Minerals: similar to white fish
Vitamin content: probably negligible

Piece of Fish cooked in batter. A piece of fish fried in batter gives approximately 200 calories. The mineral content is probably similar to that of white fish, the vitamin content is probably very poor.

Proteins	15·0 p.c.	good
Carbohydrates	7·5 p.c.	fair
Fats	15·3 p.c.	good

Crustacea

Lobster

390 calories: poor

Proteins   16·4 p.c. good
Carbohydrates 0·4 p.c. very poor
Fats   1·8 p.c. very poor
Minerals: similar to white fish
Vitamin content: probably poor

FATS

Dripping

4213 calories: very good

Proteins   —   negligible
Carbohydrates —   negligible
Fats   100 p.c. very good
Minerals: negligible
Small amount of vitamins A and D
Other vitamins: negligible

Lard

4213 calories: very good

Proteins   —   negligible
Carbohydrates —   negligible
Fats   100 p.c. very good
Minerals: negligible
Probably small amount of vitamins A and D
Other vitamins: negligible

Suet

3955 calories: very good

Proteins   1·2 p.c. very poor
Carbohydrates —   negligible
Fats   93·3 p.c. very good
Minerals: negligible
Small amount of vitamins A and D
Other vitamins: negligible

VEGETABLES
Roots

Old Potatoes

372 calories: poor

Proteins   1·9 p.c. very poor
Carbohydrates 18·1 p.c. good
Fats   —   negligible

Iron: fair
Phosphorus: poor
Calcium: fair
Vitamins A and D: negligible
Vitamin B group: poor
Vitamin C: fairly good
Vitamin E: probably negligible

New Potatoes

408 calories: poor

Proteins   1·6 p.c. very poor
Carbohydrates 20·4 p.c. very good
Fats   —   negligible

Minerals: same as old potatoes
Vitamin C: good
Other vitamins: similar to old potatoes

Potato Chips (fried)

1142 calories: fairly good

Proteins   3·8 p.c. poor
Carbohydrates 37·3 p.c. very good
Fats   9·0 p.c. fair

Iron: good
Phosphorus and calcium: fair
All vitamins: probably very poor

Beetroot

140 calories: poor

Proteins   1·2 p.c. very poor
Carbohydrates 6·2 p.c. poor
Fats   0·1 p.c. very poor

Iron, phosphorus and calcium: fair
Vitamins A and C: poor
Vitamin B group: very poor
Vitamins D and E: negligible

Carrot

204 calories: poor

Proteins   1·2 p.c. very poor
Carbohydrates   9·6 p.c. fair
Fats   0·1 p.c. negligible

Iron, phosphorus and calcium: fair
Vitamin A: fairly good
Vitamin B group: very poor
Vitamin C: fair
Vitamins D and E: negligible

Parsnip

240 calories: poor

Proteins   1·7 p.c. very poor
Carbohydrates   11·3 p.c. fair
Fats   —   negligible

Iron, phosphorus and calcium: fair
Vitamin A: fair
Vitamin B group: fair
Vitamin C: very poor
Vitamin D: negligible
Vitamin E: doubtful

Radish

82 calories: very poor

Proteins   0·8 p.c. very poor
Carbohydrates   3·3 p.c. poor
Fats   0·1 p.c. very poor

Iron: fair
Phosphorus: poor
Calcium: fair
Vitamins A, D and E: negligible
Vitamin B group: very poor
Vitamin C: very poor

Swede

172 calories: poor

Proteins   1·0 p.c. very poor
Carbohydrates   8·1 p.c. fair
Fats   0·1 p.c. very poor
Iron and phosphorus: poor
Calcium: fair

Vitamin A: poor
Vitamin B group: very poor
Vitamin C: good
Vitamins D and E: negligible

Turnip

109 calories: poor

Proteins	1·2 p.c. very poor
Carbohydrates	4·4 p.c. poor
Fats	0·1 p.c. very poor

Iron: poor
Phosphorus and calcium: fair
Vitamins: similar to swede

Green Vegetables

Cabbage

190 calories: poor

Proteins	1·3 p.c. very poor
Carbohydrates	8·7 p.c. fair
Fats	0·1 p.c. very poor

Iron: good
Phosphorus: poor
Calcium: fair
Vitamin A: good
Vitamin B group: fair
Vitamin C: fair
Vitamin D: negligible
Vitamin E: probably fair

Red Cabbage

186 calories: poor

Proteins	2·5 p.c. poor
Carbohydrates	7·1 p.c. fair
Fats	0·2 p.c. very poor

Minerals and vitamins: similar to cabbage

Savoy Cabbage

154 calories: poor

Proteins	2·3 p.c. poor
Carbohydrates	5·5 p.c. poor
Fats	0·2 p.c. very poor

Minerals and vitamins: similar to cabbage

Kale

254 calories: very poor

Proteins	3·9 p.c. poor
Carbohydrates	8·8 p.c. fair
Fats	o·4 p.c. very poor

Iron, phosphorus and calcium: good
Vitamins: similar to cabbage

Lettuce

41 calories: very poor

Proteins	o·7 p.c. very poor
Carbohydrates	1·2 p.c. very poor
Fats	o·1 p.c. very poor

Minerals: similar to cabbage
Vitamin A: fairly good in green parts
Vitamin B group: very poor
Vitamin C: good in green parts
Vitamin D: negligible
Vitamin E: good in green parts

It is important to note that the greener the lettuce the more
minerals and vitamins it contains. The pale green or white
parts of the lettuce contain very little of either of these
essential constituents. The same applies for other leafy
vegetables, such as the cabbage and kale.

Parsley

299 calories: poor

Proteins	3·1 p.c. poor
Carbohydrates	11·7 p.c. fair
Fats	o·6 p.c. very poor

Iron: very good
Phosphorus and calcium: fair
Vitamin A: fairly good in green parts
Vitamin B group: fair
Vitamin C: very good
Vitamin D: negligible
Vitamin E: probably fair

Spinach

95 calories: very poor

Proteins	1·8 p.c. very poor
Carbohydrates	2·9 p.c. poor
Fats	o·2 p.c. very poor

Iron: very good
Phosphorus and calcium: fair
Vitamin A: fair
Vitamin B group: very poor
Vitamin C: good
Vitamin D: negligible
Vitamin E: probably fair

Brussels Sprouts

181 calories: poor

Proteins	3·6 p.c. poor
Carbohydrates	5·7 p.c. poor
Fats	0·2 p.c. very poor

Iron: fair
Phosphorus: very good
Calcium: fair
Vitamin A: fair
Vitamin B group: very poor
Vitamin C: good
Vitamin D: negligible
Vitamin E: probably fairly good

Watercress

91 calories: very poor

Proteins	1·3 p.c. very poor
Carbohydrates	3·0 p.c. poor
Fats	0·3 p.c. very poor

Iron: very good
Phosphorus: very poor
Calcium: very good
Vitamin A: poor
Vitamin B group: fair
Vitamin C: very good
Vitamin D: negligible
Vitamin E: probably good

Legumes

Broad Beans in pods

494 calories: poor

Proteins	6·9 p.c. fair
Carbohydrates	19·0 p.c. good
Fats	0·3 p.c. very poor

Broad Beans in pods (*cont.*)

Iron: good
Phosphorus: very good
Calcium: fair
Vitamin A: poor
Vitamin B group: poor
Vitamin C: fairly good
Vitamin D: negligible
Vitamin E: probably fair

Scarlet Runner (French) Beans

127 calories: poor

Proteins   1·9 p.c. very poor
Carbohydrates   4·8 p.c. poor
Fats   0·1 p.c. very poor

Iron, phosphorus and calcium: fair
Vitamin A: poor
Vitamin B group: very poor
Vitamin C: good
Vitamin D: negligible
Vitamin E: probably fair

Peas in pod

403 calories: poor

Proteins   5·4 p.c. poor
Carbohydrates   15·2 p.c. good
Fats   0·5 p.c. poor

Iron and phosphorus: very good
Calcium: fair
Vitamin A: poor
Vitamin B group: poor
Vitamin C: very good
Vitamin D: negligible
Vitamin E: probably fair

Dried Legumes

Butter Beans

1345 calories: fairly good

Proteins   19·2 p.c. good
Carbohydrates   49·8 p.c. very good
Fats   1·5 p.c. very poor

Iron and phosphorus: very good
Calcium: good
Vitamin A: poor
Vitamin B group: fair
Other vitamins: negligible

Haricot Beans

1305 calories: fairly good

Proteins   21·4 p.c. very good
Carbohydrates   45·5 p.c. very good
Fats   1·5 p.c. very poor

Minerals and vitamins: similar to butter beans

Lentil

1576 calories: good

Proteins   20·1 p.c. very good
Carbohydrates   63·9 p.c. very good
Fats   0·4 p.c. very poor

Minerals and vitamins: similar to butter beans

Split Peas

1604 calories: good

Proteins   22·1 p.c. very good
Carbohydrates   62·4 p.c. very good
Fats   0·8 p.c. very poor

Minerals and vitamins: similar to butter beans

Peasemeal

1672 calories: good

Proteins   24·7 p.c. very good
Carbohydrates   61·1 p.c. very good
Fats   1·9 p.c. very poor

Minerals and vitamins: similar to butter beans

Other Vegetables

Asparagus

140 calories: poor

Proteins   3·1 p.c. poor
Carbohydrates   4·3 p.c. poor
Fats   0·1 p.c. very poor

Asparagus (*cont.*)   *Iron: good*
Phosphorus and calcium: fair
Vitamin A: poor
Vitamin B group: poor
Vitamin C: fair
Vitamins D and E: negligible

Cauliflower   **154 calories: poor**

Proteins	1·9 p.c. very poor
Carbohydrates	5·9 p.c. poor
Fats	0·2 p.c. negligible

Iron: good
Phosphorus and calcium: fair
Vitamin A: poor
Vitamin B group: poor
Vitamin C: fair
Vitamin D: negligible
Vitamin E: probably poor

Celery   **86 calories: very poor**

Proteins	0·6 p.c. very poor
Carbohydrates	3·8 p.c. poor
Fats	0·1 p.c. very poor

Iron and phosphorus: poor
Calcium: fair
Vitamin A: very poor
Vitamin B group: very poor
Vitamin C: fair (in green parts)
Vitamin D: negligible
Vitamin E: probably poor

Chives   **204 calories: poor**

Proteins	3·8 p.c. poor
Carbohydrates	5·8 p.c. poor
Fats	0·6 p.c. very poor

Iron, phosphorus and calcium: fair
Vitamin A: poor
Vitamin B group: poor
Vitamin C: fairly good in green parts
Vitamin D: negligible
Vitamin E: probably poor

Cucumber

54 calories: very poor

Proteins	0·6 p.c. very poor
Carbohydrates	2·0 p.c. very poor
Fats	0·1 p.c. very poor

Iron and phosphorus: fair
Calcium: poor
Vitamins A and D: negligible
Vitamin B group: very poor
Vitamin C: fair
Vitamin E: probably very poor

Leek

227 calories: poor

Proteins	2·5 p.c. very poor
Carbohydrates	9·3 p.c. fair
Fats	0·1 p.c. very poor

Iron: fair
Phosphorus: very poor
Calcium: fair
Vitamin A: fair in green parts
Vitamin B group: fair
Vitamin C: fairly good
Vitamin D: negligible
Vitamin E: probably poor

Marrow (green)

104 calories: poor

Proteins	0·4 p.c. very poor
Carbohydrates	5·2 p.c. poor
Fats	0·1 p.c. very poor

Iron, phosphorus and calcium: very poor
Vitamin A: very poor
Vitamin B group: very poor
Vitamin C: fair
Vitamins D and E: negligible

Marrow (yellow)

45 calories: very poor

Proteins	0·2 p.c. very poor
Carbohydrates	2·2 p.c. very poor
Fats	0·1 p.c. very poor

Minerals and vitamins: similar to green marrow

Mushroom

204 calories: poor

Proteins	3·5 p.c.	poor
Carbohydrates	6·8 p.c.	fair
Fats	0·4 p.c.	very poor

Iron: fair
Phosphorus: good
Calcium: very poor
Vitamin A: poor
Vitamin B group: poor
Vitamin C: very poor
Vitamin D: fair
Vitamin E: probably very poor

Onion (Spanish)

91 calories: very poor

Proteins	0·6 p.c.	very poor
Carbohydrates	4·1 p.c.	very poor
Fats	—	negligible

Iron, phosphorus and calcium: fair
Vitamin A: negligible
Vitamin B group: very poor
Vitamin C: poor
Vitamin D: very poor
Vitamin E: probably very poor

Pickles (sour mixed)

18 calories: very poor

Proteins	0·5 p.c.	very poor
Carbohydrates	0·3 p.c.	very poor
Fats	—	negligible

Minerals: negligible
Vitamins: probably negligible; may be a little vitamin C

Shallots (bulbs)

353 calories: poor

Proteins	1·2 p.c.	very poor
Carbohydrates	17·3 p.c.	good*
Fats	0·2 p.c.	very poor

Iron, phosphorus and calcium: fair
Vitamins: probably similar to Spanish onion

* This is unusually high for shallots; other authorities give a much lower figure for carbohydrates.

Canned Vegetables

Baked Beans

593 calories: fair

Proteins   7·5 p.c. fair
Carbohydrates   20·0 p.c. very good
Fats   2·0 p.c. very poor

Iron and phosphorus: very good
Calcium: fair
Vitamins: probably similar to butter beans

Beetroot (strained)

172 calories: poor

Proteins   1·5 p.c. very poor
Carbohydrates   7·5 p.c. fair
Fats   0·1 p.c. very poor

Iron: very good
Phosphorus: fair
Calcium: poor
Vitamins: probably negligible

Carrots

118 calories: poor

Proteins   0·9 p.c. very poor
Carbohydrates   5·2 p.c. poor
Fats   0·1 p.c. very poor

Iron: good
Phosphorus and calcium: fair
Vitamins: probably similar to, but slightly
  less than, raw carrots

Peas

254 calories: poor

Proteins   3·4 p.c. poor
Carbohydrates   10·0 p.c. fair
Fats   0·1 p.c. very poor

Iron and phosphorus: very good
Calcium: fair
Vitamins: probably similar to, but slightly
  less than, fresh green peas

FRUIT

Apple

213 calories: poor

Proteins	0·3 p.c. very poor
Carbohydrates	10·8 p.c. fairly good
Fats	0·2 p.c. very poor

Iron: fair
Phosphorus and calcium: very poor
Vitamin A: poor
Vitamin B group: very poor
Vitamin C: fairly good
Vitamin D: negligible
Vitamin E: probably negligible

Apricot

267 calories: poor

Proteins	1·1 p.c. very poor
Carbohydrates	13·4 p.c. fairly good
Fats	— negligible

Iron: fair
Phosphorus and calcium: very poor
Vitamin A: fairly good
Vitamin B group: very poor
Vitamin C: fair
Vitamins D and E: negligible

Banana

489 calories: fair

Proteins	1·2 p.c. very poor
Carbohydrates	24·9 p.c. very good
Fats	0·1 p.c. very poor

Iron: fair
Phosphorus: poor
Calcium: very poor
Vitamin A: fair
Vitamin B group: very poor
Vitamin C: fair
Vitamins D and E: negligible

Cherry

258 calories: poor

Proteins	0·9 p.c.	very poor
Carbohydrates	12·8 p.c.	fairly good
Fats	0·1 p.c.	very poor

Iron: fair
Phosphorus: poor
Calcium: very poor
Vitamin A: poor
Vitamin B group: poor
Vitamin C: fairly good
Vitamins D and E: negligible

Currant (black)

150 calories: poor

Proteins	1·2 p.c.	very poor
Carbohydrates	6·7 p.c.	fair
Fats	0·1 p.c.	negligible

Iron, phosphorus and calcium: fair
Vitamin A: poor
Vitamin B group: very poor
Vitamin C: good
Vitamins D and E: negligible

Currant (red)

104 calories: poor

Proteins	1·1 p.c.	very poor
Carbohydrates	4·4 p.c.	poor
Fats	—	negligible

Minerals and vitamins: similar to black currants

Gooseberry

127 calories: poor

Proteins	0·7 p.c.	very poor
Carbohydrates	6·0 p.c.	fair
Fats	0·1 p.c.	very poor

Minerals and vitamins: probably similar to black currants

Grape Fruit

122 calories: poor

Proteins   0·6 p.c. very poor
Carbohydrates   5·7 p.c. poor
Fats   0·1 p.c. very poor

Iron: poor
Phosphorus and calcium: fair
Vitamin A: very poor
Vitamin B group: very poor
Vitamin C: very good
Vitamins D and E: negligible

Grape

272 calories: poor

Proteins   0·6 p.c. very poor
Carbohydrates   13·9 p.c. fairly good
Fats   0·1 p.c. very poor

Iron, phosphorus and calcium: poor
Vitamin A: very poor
Vitamin B group: very poor
Vitamin C: poor
Vitamins D and E: negligible

Lemon

86 calories: very poor

Proteins   0·5 p.c. very poor
Carbohydrates   3·1 p.c. very poor
Fats   0·5 p.c. very poor

Iron, phosphorus and calcium: fair
Vitamin A: poor
Vitamin B group: poor
Vitamin C: very good
Vitamins D and E: negligible

Melon

100 calories: poor

Proteins   0·5 p.c. very poor
Carbohydrates   4·6 p.c. poor
Fats   0·1 p.c. very poor

Minerals and vitamins: poor

Orange

181 calories: poor

Proteins	0·8 p.c.	very poor
Carbohydrates	8·8 p.c.	fair
Fats	0·1 p.c.	very poor

Iron: fairly good
Phosphorus: poor
Calcium: fair
Vitamin A: fairly good
Vitamin B group: poor
Vitamin C: very good
Vitamins D and E: negligible

Peach

109 calories: poor

Proteins	0·8 p.c.	very poor
Carbohydrates	4·7 p.c.	poor
Fats	0·1 p.c.	very poor

Iron: fair
Phosphorus: poor
Calcium: very poor
Vitamins A and C: fair
Vitamin B group: very poor
Vitamins D and E: negligible

Pear

168 calories: poor

Proteins	0·3 p.c.	very poor
Carbohydrates	8·4 p.c.	fair
Fats	0·1 p.c.	very poor

Iron, phosphorus and calcium: poor
Vitamin A: very poor
Vitamin B group: very poor
Vitamin C: fair
Vitamins D and E: negligible

Plum

172 calories: poor

Proteins	0·5 p.c.	very poor
Carbohydrates	8·2 p.c.	fair
Fats	0·3 p.c.	very poor

Plum (*cont.*)

Iron, phosphorus and calcium: poor
Vitamin A: very poor
Vitamin B group: very poor
Vitamin C: fairly good
Vitamins D and E: negligible

Raspberry

150 calories: poor

Proteins	1·2 p.c. very poor
Carbohydrates	6·6 p.c. fair
Fats	0·1 p.c. very poor

Iron, phosphorus and calcium: fair
Vitamin A: poor
Vitamin B group: very poor
Vitamin C: good
Vitamins D and E: negligible

Rhubarb

82 calories: very poor

Proteins	0·6 p.c. very poor
Carbohydrates	3·6 p.c. poor
Fats	0·1 p.c. very poor

Iron and phosphorus: poor
Calcium: fair
Vitamin A: very poor
Vitamin B group: very poor
Vitamin C: fair
Vitamins D and E: negligible

Strawberry

168 calories: very poor

Proteins	0·7 p.c. very poor
Carbohydrates	8·2 p.c. fair
Fats	0·1 p.c. very poor

Iron and phosphorus: poor
Calcium: fair
Vitamin A: very poor
Vitamin B group: very poor
Vitamin C: very good
Vitamins D and E: negligible

Tomato   **100 calories: poor**

Proteins   o·7 p.c. very poor
Carbohydrates   4·5 p.c. fair
Fats   o·1 p.c. very poor

Iron: poor
Phosphorus: fair
Calcium: poor
Vitamin A: fairly good
Vitamin B group: poor
Vitamin C: very good
Vitamins D and E: negligible

Canned Fruits

Most reliable brands of canned fruits are now canned by methods
which preserve the vitamin content. There is usually only
a small loss of vitamins compared with amount in fresh
fruit. In almost every case the vitamin value will be greater
than that of the same fruit boiled or stewed in an open
saucepan on the stove.

Apricot   **476 calories: poor**

Proteins   o·7 p.c. very poor
Carbohydrates   25·0 p.c. very good
Fats   —   negligible
Minerals and vitamins: similar to fresh
    apricots

Cherry   **381 calories: poor**

Proteins   o·6 p.c. very poor
Carbohydrates   20·0 p.c. very good
Fats   —   negligible
Minerals and vitamins: probably similar to
    fresh cherries

Fig   **797 calories: fair**

Proteins   1·2 p.c. very poor
Carbohydrates   40·9 p.c. very good
Fats   o·3 p.c. very poor

Fig (*cont.*)   *Iron: very good*
Phosphorus and calcium: good
Vitamin A: poor
Vitamin B group: very poor
Vitamin C: fair
Vitamins D and E: negligible

Fruit Salad   **381 calories: poor**

Proteins   0·5 p.c. very poor
Carbohydrates   20·0 p.c. very good
Fats   —   negligible
Iron: fair
Phosphorus and calcium: very poor
Vitamins A and C: fair
Vitamin B group: very poor
Vitamins D and E: negligible

Gooseberry   **308 calories: poor**

Proteins   0·7 p.c. very poor
Carbohydrates   16·0 p.c. good
Fats   0·1 p.c. very poor
Minerals and vitamins: probably similar to
  fresh gooseberries

Orange   **281 calories: poor**

Proteins   0·5 p.c. very poor
Carbohydrates   14·7 p.c. fairly good
Fats   —   negligible
Minerals and vitamins: probably similar to
  fresh oranges

Peach   **476 calories: poor**

Proteins   0·5 p.c. very poor
Carbohydrates   25·0 p.c. very good
Fats   —   negligible
Minerals and vitamins: probably similar to
  fresh peaches

Pear   **381 calories: poor**

Proteins   0·4 p.c. very poor
Carbohydrates   20·0 p.c. very good
Fats   —   negligible

Minerals and vitamins: probably similar to fresh pears

Pineapple   **471 calories: poor**

Proteins   0·4 p.c. very poor
Carbohydrates   25·0 p.c. very good
Fats   —   negligible

Iron: fair
Phosphorus and calcium: very poor
Vitamin A: fair
Vitamin B group: very poor
Vitamin C: good
Vitamins D and E: negligible

Plum   **476 calories: poor**

Proteins   0·5 p.c. very poor
Carbohydrates   25·0 p.c. very good
Fats   —   negligible

Minerals and vitamins: probably similar to fresh plums

Prune   **426 calories: poor**

Proteins   0·5 p.c. very poor
Carbohydrates   22·3 p.c. very good
Fats   0·1 p.c. negligible

Iron: very good
Phosphorus and calcium: good
Vitamin A: poor
Vitamin B group: poor
Vitamins C, D and E: negligible

Tomato   **104 calories: poor**

Proteins   1·2 p.c. very poor
Carbohydrates   4·0 p.c. poor
Fats   0·2 p.c. very poor

Minerals and vitamins: probably similar to fresh tomatoes

Tomato purée   **199 calories: poor**

Proteins	2·1 p.c.	very poor
Carbohydrates	8·2 p.c.	fair
Fats	0·2 p.c.	very poor

Iron: very good
Phosphorus and calcium: poor
Vitamins: probably similar to fresh tomatoes

Dried Fruits

Apple   **1133 calories: fairly good**

Proteins	0·9 p.c.	very poor
Carbohydrates	58·7 p.c.	very good
Fats	0·6 p.c.	very poor

Iron: very good
Phosphorus and calcium: fair
Vitamins: negligible

Apricot   **1037 calories: fairly good**

Proteins	5·5 p.c.	poor
Carbohydrates	49·6 p.c.	very good
Fats	0·3 p.c.	very poor

Iron and phosphorus: very good
Calcium: good
Vitamin A: poor
Other vitamins: negligible

Currant   **825 calories: fair**

Proteins	1·7 p.c.	very poor
Carbohydrates	42·0 p.c.	very good
Fats	0·3 p.c.	very poor

Iron: good
Phosphorus: very good
Calcium: good
Vitamins: negligible

Fig   **1110 calories: fairly good**

Proteins	2·0 p.c.	very poor
Carbohydrates	56·5 p.c.	very good
Fats	0·5 p.c.	very poor

Iron, phosphorus and calcium: very good
Vitamins: negligible

Fruit Salad

924 calories: fairly good

Proteins   3·1 p.c. poor
Carbohydrates 45·7 p.c. very good
Fats   0·4 p.c. very poor

Iron and phosphorus: very good
Calcium: good
Vitamins: negligible

Peach

530 calories: fair

Proteins   4·0 p.c. poor
Carbohydrates 23·5 p.c. very good
Fats   0·5 p.c. very poor

Iron and phosphorus: very good
Calcium: good
Vitamins: negligible

Prune

820 calories: fair

Proteins   3·0 p.c. poor
Carbohydrates 40·4 p.c. very good
Fats   0·3 p.c. very poor

Iron and phosphorus: very good
Calcium: fairly good
Vitamins: negligible

Raisin

1191 calories: fairly good

Proteins   2·2 p.c. very poor
Carbohydrates 61·2 p.c. very good
Fats   0·3 p.c. very poor

Iron and phosphorus: very good
Calcium: fairly good
Vitamins: negligible

Sultana

1250 calories: fairly good

Proteins   1·7 p.c. very poor
Carbohydrates 64·9 p.c. very good
Fats   0·3 p.c. very poor

Sultana (*cont.*) *Iron: very good*
Phosphorus: good
Calcium: fairly good
Vitamins: negligible

Candied Peel   **1250 calories: fairly good**

Proteins   0·3 p.c. very poor
Carbohydrates 66·8 p.c. very good
Fats   0·1 p.c. very poor
Minerals and vitamins: negligible

Date   **1278 calories: fairly good**

Proteins   1·6 p.c. very poor
Carbohydrates 66·8 p.c. very good
Fats   0·2 p.c. very poor
Iron: good
Phosphorus and calcium: fair
Vitamins: negligible

NUTS

Almond   **2976 calories: very good**

Proteins   18·8 p.c. good
Carbohydrates 16·0 p.c. good
Fats   55·3 p.c. very good
Iron, phosphorus and calcium: very good
Vitamin A: poor
Vitamin B group: good
Vitamin C: negligible
Vitamin D: very poor
Vitamin E: probably fairly good

Cocoanut (desiccated) **2632 calories: very good**

Proteins   4·3 p.c. poor
Carbohydrates 44·5 p.c. very good
Fats   41·0 p.c. very good
Iron and phosphorus: very good
Calcium: fair
Vitamins: probably all very poor

Walnut (dried)   **3311 calories: very good**

Proteins   13·6 p.c. fairly good
Carbohydrates 14·0 p.c. fairly good
Fats   66·4 p.c. very good

Iron: fairly good
Phosphorus: very good
Calcium: fair
Vitamin A: probably fair
Vitamin B group: probably fair
Other vitamins: probably negligible

VITAMIN CONCENTRATES

Bemax   **1853 calories: good**

Proteins   34·0 p.c. very good
Carbohydrates 46·5 p.c. very good
Fats   8·5 p.c. fair

Iron and phosphorus: very good
Calcium: fair
Vitamin A: fair
Vitamin B group: very good
Vitamins C and D: negligible
Vitamin E: very good

Cod Liver Oil   **4213 calories: very good**

Proteins   —   negligible
Carbohydrates —   negligible
Fats   100·0 p.c. very good

Minerals: negligible
Vitamins A and D: very good
Other vitamins: negligible

Marmite   **calories probably negligible**

Proteins   Doubtful
Carbohydrates Probably negligible
Fats   Doubtful

Iron, phosphorus and calcium: very good
Vitamin B group: very good
Other vitamins: probably negligible

CEREALS

Wholemeal Flour

1635 calories: good

Proteins   11·7 p.c. fairly good
Carbohydrates 71·9 p.c. very good
Fats   2·0 p.c. very poor
Iron, phosphorus and calcium: fair
Vitamin A: fairly good
Vitamin B group: very good
Vitamins C and D: negligible
Vitamin E: very good

White Flour

1663 calories: good

Proteins   12·6 p.c. fairly good
Carbohydrates 73·1 p.c. very good
Fats   1·7 p.c. very poor
Iron, phosphorus and calcium: poor
Vitamins: all very poor

White Bread

1182 calories: fairly good

Proteins   8·9 p.c. fair
Carbohydrates 54·0 p.c. very good
Fats   0·3 p.c. very poor
Minerals and vitamins: similar to white flour

Wholemeal Bread

1155 calories: fairly good

Proteins   8·2 p.c. fair
Carbohydrates 52·9 p.c. very good
Fats   0·5 p.c. very poor
Minerals and vitamins: similar to wholemeal flour

Buns (plain)

1749 calories: good

Proteins   12·5 p.c. fairly good
Carbohydrates 61·0 p.c. very good
Fats   9·1 p.c. fair
Minerals and vitamins: similar to white flour

Buns (cream)

2088 calories: very good

Proteins   11·2 p.c. fair
Carbohydrates 53·2 p.c. very good
Fats   21·2 p.c. very good

Iron, phosphorus and calcium: fair
Vitamin A: fairly good
Vitamin B group: poor
Vitamins C and E: negligible
Vitamin D: fair

Doughnuts

1993 calories: good

Proteins   6·7 p.c. fair
Carbohydrates 53·1 p.c. very good
Fats   21·0 p.c. very good

Minerals: similar to plain buns
Vitamins: all poor

Scones (white)

1336 calories: fairly good

Proteins   12·2 p.c. fairly good
Carbohydrates 58·4 p.c. very good
Fats   0·6 p.c. poor

Minerals and vitamins: similar to white
  flour

Scones (wholemeal)

1418 calories: fairly good

Proteins   11·9 p.c. fairly good
Carbohydrates 60·7 p.c. very good
Fats   1·6 p.c. very poor

Minerals and vitamins: similar to whole-
  meal flour

BISCUITS

Digestive Biscuits
  (wholemeal)

2356 calories: very good

Proteins   5·6 p.c. poor
Carbohydrates 64·6 p.c. very good
Fats   25·0 p.c. very good

Digestive Biscuits
(*cont.*)

Iron and phosphorus: very good
Calcium: fair
Vitamin A: poor
Vitamin B group: fairly good
Vitamins C and D: negligible
Vitamin E: fairly good

Gingersnaps

1894 calories: good

Proteins   6·5 p.c. fair
Carbohydrates 76·0 p.c. very good
Fats   8·6 p.c. fair

Iron and phosphorus: fair
Calcium: poor
Vitamins: probably all very poor

Plain Biscuits

1966 calories: good

Proteins   9·7 p.c. fair
Carbohydrates 68·3 p.c. very good
Fats   12·3 p.c. fairly good

Iron and phosphorus: fair
Calcium: poor
Vitamins: probably all very poor

Sweet Biscuits

2360 calories: very good

Proteins   6·4 p.c. fair
Carbohydrates 64·2 p.c. very good
Fats   24·9 p.c. very good

Minerals and vitamins: similar to plain
biscuits

Shortbread

2460 calories: very good

Proteins   5·7 p.c. poor
Carbohydrates 61·7 p.c. very good
Fats   28·7 p.c. very good

Minerals and vitamins: similar to plain
biscuits

CAKES

Genoa

1744 calories: good
Proteins   4·4 p.c. poor
Carbohydrates 60·1 p.c. very good
Fats   13·0 p.c. fairly good
Minerals: similar to plain biscuits
Vitamins A and D: probably poor
Other vitamins: negligible

Fondant covered Sandwich

2097 calories: very good
Proteins   6·1 p.c. fair
Carbohydrates 63·6 p.c. very good
Fats   19·1 p.c. good
Iron, phosphorus and calcium: fair
If cream is used in the fondant, vitamins A
  and D will probably be fairly good
Other vitamins: negligible

Jam Tart

1835 calories: good
Proteins   4·2 p.c. poor
Carbohydrates 63·9 p.c. very good
Fats   13·5 p.c. fairly good
Minerals and vitamins: similar to plain
  biscuits

Madeira Cake

1685 calories: good
Proteins   5·5 p.c. poor
Carbohydrates 52·7 p.c. very good
Fats   14·3 p.c. fairly good
Minerals and vitamins: similar to plain
  biscuits

Meringue (cases only, without filling)

1798 calories: good
Proteins   5·6 p.c. poor
Carbohydrates 91·2 p.c. very good
Fats   —   negligible
Minerals and vitamins: similar to plain
  biscuits

Sponge Sandwich   **1558 calories: good**

Proteins   9·1 p.c. fair
Carbohydrates 62·7 p.c. very good
Fats   5·3 p.c. poor

Minerals and vitamins: similar to plain biscuits, unless cream is used as a filling, in which case there will be some vitamins A and D present

Swiss Roll   **1499 calories: fairly good**

Proteins   7·5 p.c. fair
Carbohydrates 62·6 p.c. very good
Fats   4·7 p.c. poor

Minerals and vitamins: similar to plain biscuits

CEREAL PRODUCTS (INCLUDING BREAKFAST FOODS)

Wheat

Puffed Wheat   **1735 calories: good**

Proteins   16·2 p.c. good
Carbohydrates 73·2 p.c. very good
Fats   1·8 p.c. very poor

Iron, phosphorus and calcium: fairly good
Vitamin A: fair
Vitamin B group: good
Vitamins C and D: negligible
Vitamin E: fairly good

Shredded Wheat   **1672 calories: good**

Proteins   9·8 p.c. fair
Carbohydrates 77·7 p.c. very good
Fats   1·1 p.c. very poor

Minerals and vitamins: similar to puffed wheat

Other Cereals

Barley Pearl   **1649 calories: good**

Proteins   7·0 p.c. fair
Carbohydrates 79·9 p.c. very good
Fats   0·8 p.c. very poor

Iron and phosphorus: fairly good
Calcium: poor
Vitamins: poor

Cornflour   **1644 calories: good**

Proteins   0·8 p.c. very poor
Carbohydrates 87·6 p.c. very good
Fats   0·1 p.c. very poor
Minerals: poor
Vitamins: negligible

Cornflakes. These are composed of maize which has been cooked, mixed with malt honey, and rolled, dried, and baked. Wheat flakes are made from wheat grains in the same way.

1767 calories: good

Proteins   7·4 p.c. fair
Carbohydrates 87·4 p.c. very good
Fats   0·2 p.c. very poor
Iron: fairly good
Phosphorus and calcium: fair
Vitamin A: fairly good
Vitamin B group: fair (nicotinic acid very poor)
Vitamins C and D: negligible
Vitamin E: probably fair

Oatmeal   **1885 calories: good**

Proteins   11·9 p.c. fairly good
Carbohydrates 70·0 p.c. very good
Fats   8·6 p.c. fair
Iron and phosphorus: fairly good
Calcium: fair
Vitamin A: fair
Vitamin B group: fairly good
Vitamins C and D: negligible
Vitamin E: probably fairly good

Oat Cakes

1971 calories: good

Proteins	11·8 p.c. fairly good
Carbohydrates	69·0 p.c. very good
Fats	11·1 p.c. fairly good

Minerals and vitamins: similar to oatmeal

Rice (polished; whole or ground)

1649 calories: good

Proteins	6·4 p.c. fair
Carbohydrates	81·6 p.c. very good
Fats	0·3 p.c. very poor

Iron and phosphorus: fair
Calcium: very poor
Vitamins: all very poor

Rice (unpolished)

1635 calories: good

Proteins	6·8 p.c. fair
Carbohydrates	80·0 p.c. very good
Fats	0·6 p.c. very poor

Iron and phosphorus: very good
Calcium: fairly good
Vitamin A: fair
Vitamin B group: good
Vitamins C and D: negligible
Vitamin E: probably fairly good

Farinaceous Foods

Arrowroot. This is a starchy powder obtained by pulping the rhizomes (root-like organs) of a plant which grows in the West Indies, and washing the pulp with water. The water carries away the starch. It is subsequently evaporated and the starch obtained as a fine white powder. Farina—English arrowroot—is usually prepared from potatoes or maize.

1608 calories: good

Proteins	0·2 p.c. very poor
Carbohydrates	86·4 p.c. very good
Fats	— negligible

Iron and phosphorus: fair
Calcium: poor
Vitamins: negligible

Custard Powder. This is usually made of powdered starch coloured by a vegetable dye (turmeric). Custards made from these powders are obviously useful only as a supply of energy, and they have no protective value at all unless the custard is made from milk. Because they resemble true custards (made with milk and eggs) in appearance many people consider them similar in nutritive value, but this is far from the truth.

1581 calories: good

Proteins	0·3 p.c.	very poor
Carbohydrates	84·6 p.c.	very good
Fats	0·2 p.c.	very poor

Minerals: very poor
Vitamins: negligible

Farola. Is a starchy preparation made from the wheat grain.

1622 calories: good

Proteins	12·5 p.c.	fairly good
Carbohydrates	72·2 p.c.	very good
Fats	1·2 p.c.	very poor

Iron and phosphorus: fair
Calcium: good
Vitamin A: very poor
Vitamin B group: poor
Other vitamins: probably negligible

Force. This is composed of whole wheat grains treated in the same way as in the manufacture of cornflakes.

1735 calories: good

Proteins	11·4 p.c.	fairly good
Carbohydrates	79·5 p.c.	very good
Fats	1·1 p.c.	very poor

Iron and phosphorus: very good
Calcium: poor
Vitamin A: fair
Vitamin B group: very good
Vitamins C and D: negligible
Vitamin E: good

Macaroni   **1654 calories: good**

Proteins   11·7 p.c. fairly good
Carbohydrates 76·8 p.c. very good
Fats   0·2 p.c. very poor
Iron and phosphorus: fairly good
Calcium: poor
Vitamins: very poor

Sago. This is obtained from the sago palm. The trees are felled and split open, and the starch is washed out with water which is subsequently evaporated.

1635 calories: good

Proteins   0·2 p.c. very poor
Carbohydrates 87·7 p.c. very good
Fats   —   negligible
Iron: good
Phosphorus: fair
Calcium: poor
Vitamins: negligible

Semolina. This is derived from the starchy part of certain hard wheats.

1658 calories: good

Proteins   11·4 p.c. fairly good
Carbohydrates 75·3 p.c. very good
Fats   1·1 p.c. very poor
Iron and phosphorus: fair
Calcium: good
Vitamins: very poor

Tapioca. Is obtained from the roots of the South American cassava plant. The roots are grated and washed with water and the tapioca collected after the water has been evaporated.

1635 calories: good

Proteins   0·2 p.c. very poor
Carbohydrates 87·7 p.c. very good
Fats   0·1 p.c. very poor

Iron: good
Phosphorus: fairly good
Calcium: poor
Vitamins: negligible

MISCELLANEOUS

Fruit Pie   **1839 calories: good**

Proteins   2·6 p.c. very poor
Carbohydrates 61·5 p.c. very good
Fats   15·4 p.c. good
Minerals: very poor
Vitamins: negligible

Jelly Crystals   **1780 calories: very good**

Proteins   21·6 p.c. very good*
Carbohydrates 74·0 p.c. very good
Fats   0·1 p.c. very poor
Minerals: very poor
Vitamins: none

* Gelatin, which is an incomplete protein.

Lemon Curd. This, if made, as it should be, with butter, eggs, sugar and lemon juice, is an extremely wholesome and nutritious food; but as usually made commercially, it contains chiefly starch, artificially coloured and flavoured, which makes it a supplier of energy and very little else. The analysis shown below is of the latter type.

1422 calories: fairly good

Proteins   2·1 p.c. very poor
Carbohydrates 61·4 p.c. very good
Fats   5·8 p.c. poor
Iron: fair
Phosphorus and calcium: poor
Vitamins: negligible

Pudding Mixture   **1626 calories: good**

Proteins   5·1 p.c. poor
Carbohydrates 81·8 p.c. very good
Fats   0·3 p.c. very poor

Pudding Mixture (*cont.*)

> *Iron: very poor*
> *Phosphorus: fair*
> *Calcium: very poor*
> *Vitamins: negligible*

Sugar and Sugar Products

Golden Syrup. Obtained as a by-product in the manufacture of crystallised sugar.

> **1427 calories: fairly good**
>
> Proteins   0·3 p.c. very poor
> Carbohydrates 76·4 p.c. very good
> Fats   —   negligible
>
> *Iron: poor*
> *Phosphorus and calcium: very poor*
> *Vitamins: negligible*

Sugar (white)

> **1857 calories: good**
>
> Proteins   —   none
> Carbohydrates 100·0 p.c. very good
> Fats   —   none
>
> *Minerals and vitamins: none*

Sugar (brown)

> **1821 calories: good**
>
> Proteins   —   none
> Carbohydrates 98·0 p.c. very good
> Fats   —   none
>
> *Minerals and vitamins: none*

Treacle. Obtained as a by-product in the manufacture of crystallised sugar.

> **1142 calories: fairly good**
>
> Proteins   1·6 p.c. very poor
> Carbohydrates 59·9 p.c. very good
> Fats   —   negligible
>
> *Iron: good*
> *Phosphorus: poor*
> *Calcium: fair*
> *Vitamins: negligible*

Jam

1296 calories: fairly good

Proteins   0·3 p.c. very poor
Carbohydrates 69·4 p.c. very good
Fats   —   negligible

Minerals: poor
Vitamins: negligible

Marmalade (grape-
fruit or orange)

1296 calories: fairly good

Proteins   0·2 p.c. very poor
Carbohydrates 69·5 p.c. very good
Fats   —   negligible

Minerals: poor
Vitamins: negligible

BEVERAGES

Bournvita

1925 calories: good

Proteins   13·1 p.c. fairly good
Carbohydrates 72·3 p.c. very good
Fats   8·1 p.c. fair

Iron and phosphorus: very good
Calcium: fairly good
Vitamin B group: probably fairly good
Other vitamins: probably negligible

Chocolate

2510 calories: very good

Proteins   4·8 p.c. poor
Carbohydrates 59·9 p.c. very good
Fats   31·1 p.c. very good

Iron and phosphorus: very good
Calcium: fairly good
Vitamin B group: probably fairly good
Other vitamins: negligible

Cocoa. This is chocolate from which some of the fat has been removed.

2120 calories: very good

Proteins   18·1 p.c. good
Carbohydrates 36·0 p.c. very good
Fats   26·5 p.c. very good

Minerals and vitamins: similar to chocolate

Cocoa not only has food value but it also contains a substance called "theobromine", which is related to the drug "caffeine" found in coffee and tea and which acts as a stimulant.

Coffee. Has no food value at all. It is, however, a stimulant, owing to the fact that it contains "caffeine" and certain odoriferous principles known as "essential oils".

Ovaltine   **1862 calories: good**

Proteins   14·2 p.c. fairly good
Carbohydrates 67·9 p.c. very good
Fats   8·0 p.c. fair

Iron, phosphorus and calcium: very good
Vitamins A and D: probably fair
Vitamin B group: probably fairly good
Other vitamins: negligible

Tea. Like coffee contains "caffeine" and "essential oils" which cause it to be classed as a stimulant, and like coffee, it has no food value, apart from the milk and sugar added to it.

ALCOHOLIC BEVERAGES

There is no space in this book to discuss whether alcoholic beverages are, or are not, good for one. It is of interest, however, to mention the one published piece of important experimental work on this subject. Dr H. H. Mitchell, in the Department of Nutrition in the University of Illinois, U.S.A., found that rats fed on a diet adequate in the known nutritional factors, and to which a small daily allowance of alcohol was added, grew faster, and retained fat and nitrogen to a much greater extent than other animals which had received the same diet but not the alcohol. Alcoholic beverages can therefore be regarded as a food of a sort; some of the energy they supply comes from the alcohol and part comes from the carbohydrate which some of them contain. Alcohol itself has an energy value of approximately 7 calories per gram (one thirtieth of an ounce).

The following list gives the approximate number of calories supplied by a pint of the most common alcoholic beverages. These calories are derived from both alcohol and carbohydrate (if any).

Beer	184	calories per pint
Brandy	1351	,, ,,
Cider	143	,, ,,
Gin	1281	,, ,,
Port	763	,, ,,
Sherry (sweet)	786	,, ,,
Sherry (dry)	700	,, ,,
Whisky	1316	,, ,,

Alcoholic beverages supply chiefly energy. They contain only small amounts of minerals and no vitamins. It is important to note that the burning of alcohol to provide energy requires vitamin B_1, and the inevitable result of an alcoholic "night out" is a deficiency of this vitamin, varying according to the amount of alcohol consumed.

CONDIMENTS

Curry Powder

1804 calories: good

Proteins	13·1 p.c. fairly good
Carbohydrates	53·5 p.c. very good
Fats	13·4 p.c. fairly good

Iron: very good
Phosphorus and calcium: doubtful
Vitamin A: probably fair
Other vitamins: probably negligible

Mayonnaise

3543 calories: very good

Proteins   2·5 p.c. very poor
Carbohydrates   —   negligible
Fats   83·0 p.c. very good
Minerals: poor
Vitamins: probably poor, if mayonnaise is made with olive oil or nut oil; if made with cod liver oil, then vitamins A and D very good

Salad Cream

1404 calories: fairly good

Proteins   2·0 p.c. very poor
Carbohydrates   10·0 p.c. fair
Fats   28·0 p.c. very good
Minerals: poor
Vitamins: probably all poor

Tomato sauce

188 calories: poor

Proteins   0·4 p.c. very poor
Carbohydrates   9·0 p.c. fair
Fats   0·3 p.c. very poor
Iron: fair
Phosphorus and calcium: poor
Vitamin A: probably fair
Vitamin B group: negligible
Vitamin C: possibly fair
Vitamins D and E: negligible

Yeast (dried)

435 calories: poor

Proteins   14·1 p.c. fairly good
Carbohydrates   8·2 p.c. fair
Fats   0·5 p.c. very poor
Iron: poor
Phosphorus: fair
Calcium: poor
Vitamin A: poor
Vitamin B group: very good (best known natural source)
Vitamin C: very poor
Vitamins D and·E: poor

Sweets

Chocolate (milk)   **2605 calories: very good**
Proteins   8·0 p.c. fair
Carbohydrates   53·0 p.c. very good
Fats   35·0 p.c. very good
Iron: fairly good
Phosphorus and calcium: very good
Vitamins A, B group and D: probably fair
Other vitamins: negligible

Chocolate (plain)   **2560 calories: very good**
Proteins   4·1 p.c. poor
Carbohydrates   59·9 p.c. very good
Fats   32·5 p.c. very good
Minerals: similar to milk chocolate
Vitamins: similar to milk chocolate except
  that there is less vitamins A and D

Ginger (crystallised)   **1626 calories: good**
Proteins   —   negligible
Carbohydrates   87·5 p.c. very good
Fats   —   negligible
Minerals and vitamins: probably negligible

LIST B

The best sources of the various food elements. The substances are arranged approximately in the order of value.

Best sources of calories
Dripping, lard, olive oil, suet, margarine, butter, cheese, nuts, cream, powdered milk, fat bacon, chocolate, biscuits, sweets, breakfast foods, flour, sugar, cakes, dried peas, beans and lentils, dried fruits, honey, bread, jam, golden syrup.

Best sources of first class (animal) protein
Dried fish, smoked fish, lean meat, lean bacon, cheese, fresh fish, poultry, game, rabbits, cooked meats, tinned meats, tinned fish, powdered milk, sausages.

Best sources of second class (vegetable) protein
Nuts, dried peas, beans and lentils, soya beans, dried fruits, flours, breads, and breakfast foods.

Best sources of carbohydrate
Sugar, sweets, chocolate, dried fruits, nuts, peas, beans, lentils, biscuits, cakes, flour, bread, spaghetti, macaroni, honey, jam.

Best sources of animal fat
Cod and other fish liver oil, dripping, suet, whale oil, butter, margarine, powdered whole milk, cheese, cream, fat bacon, fat meat, eggs and fat fish.

Best sources of vegetable fat
Olive oil, nuts, chocolate, dried peas, beans and lentils.

Best sources of iron
Curry powder, ovaltine, kidney, liver, dried peas, beans and lentils, bemax, parsley, heart, breakfast foods, cocoa, marmite,

eggs, dried fruits, sardines, kale, cheese, watercress, powdered whole milk, tongue, radish, bacon, potatoes, poultry, game, fish and rabbit.

Best sources of phosphorus

Meat extracts, marmite, beemax, cheese, dried cod or whiting, powdered milk, cocoa, fresh milk, ovaltine, yeast, dried peas, beans and lentils, nuts, condensed milk, sweetbread, veal, kidney, liver, ham, pork, fish, fish roes, lobster, chocolate, eggs, bemax.

Best sources of calcium

Cheeses, all powdered milks, marmite, condensed milks, chocolate, almond, watercress, kale, fresh milk (skimmed or whole), cocoa, bournvita, dried figs, dried currants, dried peas, beans and lentils, ice cream, cream.

Best sources of vitamin A

Halibut liver oil, cod liver oil, calf liver, butter, ox liver, spinach, cheese, egg yolk, cream, tomato, carrots, watercress, cabbage, raw peas, milk (variable), runner beans, orange, parsnip, sardine, banana, currants (black or red).

Best sources of vitamin B group

Dried yeast, marmite, bemax, bran, oatmeal, whole wheat, whole barley, whole rye, brown rice, wholemeal biscuits, wholemeal breakfast foods, wholemeal breads, germ breads, dried peas, beans and lentils, cabbage, watercress, brain, sweetbread, liver, kidney, heart, egg yolk, powdered milk, parsnips, leeks, lean meat.

Best sources of vitamin C

Lemon, orange, grapefruit, spinach, rose hips, freshly sprouted seeds (e.g. mustard and cress), cabbage, strawberry, tomato, black currants, fresh pineapple, green peas, swede turnip, runner beans, potato, young carrot, peach, apple, banana, good quality tinned pineapple, strawberries, grapefruit, peas or tomato.

Best sources of vitamin D

Halibut liver oil, cod liver oil, irradiated yeast, canned salmon, herring, egg yolk, cream, butter, ox liver, milk, suet, dripping.

Best sources of vitamin E

Bemax, wholemeal bread, germ breads, wholemeal biscuits and breakfast foods, lettuce, unpolished rice, whole rye, oatmeal, green leafy vegetables.

INDEX

CAMBRIDGE: PRINTED BY W. LEWIS, M.A., AT THE UNIVERSITY PRESS

Milton Keynes UK
Ingram Content Group UK Ltd.
UKHW041520181024
449640UK00009B/77